Bioinformatics for Immunomics

Immunomics Reviews
An Official Publication of the International Immunomics Society

---

This peer-reviewed book series offers insight on immunology for 21st century. The technological revolution has borne advances in high-throughput instrumentation and information technology, initiating a renaissance for biomathematics, and biostatistics. Cross-fertilization between genomics and immunology has led to a new field called immunomics, transforming the way in which theoretical, clinical and applied immunology are practiced. Immunomics Reviews will cover integrative approaches and applications to the theory and practice of immunology and explore synergistic effects resulting from a combination of technological advances and the latest analytical tools with the traditional fields of basic and clinical immunology.

For other titles published in this series, go to
www.springer.com/series/7528

Darren R. Flower · Matthew N. Davies
Shoba Ranganathan
Editors

Bioinformatics for Immunomics

Springer

Editors
Darren R. Flower
University of Oxford
RG20 7NN
United Kingdom
darren.flower@jenner.ac.uk

Matthew N. Davies
University of London
London WC1E 7HX
United Kingdom
m.davies@mail.cryst.bbk.ac.uk

Shoba Ranganathan
Macquarie University
Sydney
Australia
shoba.ranganathan@mq.edu.au

ISBN 978-1-4419-0539-0 e-ISBN 978-1-4419-0540-6
DOI 10.1007/978-1-4419-0540-6
Springer New York Dordrecht Heidelberg London

Library of Congress Control Number: 2009927086

Printed on acid-free paper

Springer is part of Springer Science+Business Media (www.springer.com)

Contents

Contributors

P. Antal
Department of Measurement and Information Systems, Budapest
University of Technology and Economics, Magyar tudosok korutja 2.,
1117, Rm IE 423, Budapest,
Hungary

M.N. Davies
The Jenner Institute, University of Oxford, High Street, Compton,
Berkshire RG20 7NN, UK
m.davies@mail.cryst.bbk.ac.uk

Carmen M. Díez-Rivero
Facultad de Medicina, Departamento de Immunología (Microbiología I),
Universidad Complutense de Madrid, Pabellón 5º, planta 4ª,
28040 Madrid, Spain
cmdiezri@med.ucm.es

A. Falus
Department of Genetics, Cell- and Immunobiology, Semmelweis University,
Nagyvárad tér 4, Budapest 1089, Hungary
Faland@dgci.sote.hu

D.R. Flower
The Jenner Institute, University of Oxford, High Street, Compton,
Berkshire RG20 7NN, UK
darren.flower@jenner.ac.uk

Jason A. Greenbaum, Ph.D.
La Jolla Institute for Allergy & Immunology, 9420 Athena Circle,
La Jolla, CA 92037, USA
jgbaum@liai.org

G. Hajós
Department of Measurement and Information Systems, Budapest University of Technology and Economics, Magyar tudosok korutja 2., 1117 Budapest, Hungary

G. Hullám
Department of Measurement and Information Systems, Budapest University of Technology and Economics, Magyar tudosok korutja 2., 1117 Budapest, Hungary
hullam.gabor@mit.bme.hu

Javed Mohammed Khan
Department of Chemistry and Biomolecular Sciences & ARC Centre of Excellence in Bioinformatics, Macquarie University, Sydney, NSW 2109, Australia

Sneh Lata
Institute of Microbial Technology, Sector39A, Chandigarh, India
sneh@imtech.res.in

Steven G.E. Marsh
Department of Haematology, Royal Free Hospital, Pond Street, Hampstead, London NW3 2QG, UK
marsh@ebi.ac.uk

A. Millinghoffe
Department of Measurement and Information Systems, Budapest University of Technology and Economics, Magyar tudosok korutja 2., 1117 Budapest, Hungary
milli@mit.bme.hu

Bjoern Peters, Ph.D.
La Jolla Institute for Allergy & Immunology, 9420 Athena Circle, La Jolla, CA 92037, USA
bpeters@liai.org

G.P.S. Raghava
Institute of Microbial Technology, Sector39A, Chandigarh, India
raghava@imtech.res.in

Shoba Ranganathan
Department of Biochemistry, Yong Loo Lin School of Medicine, National University of Singapore, 8 Medical Drive, Singapore 117597

Pedro Reche
Facultad de Medicina, Departamento de Immunología (Microbiología I), Universidad Complutense de Madrid, Pabellón 5º, planta 4ª, 28040 Madrid, Spain, pareche@med.ucm.es

James Robinson
Anthony Nolan Research Institute, Royal Free Hospital, Pond Street, Hampstead, London NW3 2QG, UK
jrobinso@ebi.ac.uk

Alessandro Sette, Ph.D.
La Jolla Institute for Allergy & Immunology, 9420 Athena Circle, La Jolla, CA 92037, USA
alex@liai.org

Cs. Szalai
Inflammation Biology and Immunogenomics Research Group, Hungarian Academy of Sciences, Nagyvárad tér 4, Budapest 1089, Hungary

Jon Timmis, B.Sc. (Wales), Ph.D. (Wales)
Department of Computer Science, University of York, York, UK

Joo Chuan Tong
Institute for Infocomm Research, 21 Heng Mui Keng Terrace, Singapore 119613

Randi Vita, M.D.
La Jolla Institute for Allergy & Immunology, 9420 Athena Circle, La Jolla CA 92037, USA
rvita@liai.org

Laura M. Zarebski, Ph.D.
La Jolla Institute for Allergy & Immunology, 9420 Athena Circle, La Jolla, CA 92037, USA
laura@liai.org

Introduction

Like many words, the term "immunomics" equates to different ideas contingent on context. For a brief span, immunomics meant the study of the Immunome, of which there were, in turn, several different definitions. A now largely defunct meaning rendered the Immunome as the set of antigenic peptides or immunogenic proteins within a single microorganism – be that virus, bacteria, fungus, or parasite – or microbial population, or antigenic or allergenic proteins and peptides derived from the environment as a whole, containing also proteins from eukaryotic sources. However, times have changed and the meaning of immunomics has also changed. Other newer definitions of the Immunome have come to focus on the plethora of immunological receptors and accessory molecules that comprise the host immune arsenal. Today, Immunomics or immunogenomics is now most often used as a synonym for high-throughput genome-based immunology. This is the study of aspects of the immune system using high-throughput techniques within a conceptual landscape borne of both clinical and biophysical thinking.

Within an immunogenomic or immunomic framework, chapter "A Bioinformatic Platform for a Bayesian, Multiphased, Multilevel Analysis in Immunogenomics" describes a bioinformatic platform for undertaking Bayesian analysis at multilevels. How the phenotypic behaviour of the immune system emerges from the interaction of its genome-encoded components should be of paramount interest to all involved in its investigation. Saying this is one thing; but actually achieving it is quite another. Bayesian statistics can provide an insightful route to manifesting data which is both rigorous and of true utility.

Clearly, the genome, the epinome, the proteome, the glycome, the metabolome, and all the rest of the – omes and – omics that have come to dominate our current perceptions are of direct relevance to burgeoning understanding of immunology and immunological processes. Genes, proteins, carbohydrates, and lipids, as well glycoproteins and lipoproteins, together with the peptides and small molecules too, all take part in an extraordinary range of interactions that manifest themselves as our immune response to pathogen challenge. It is clear, if still largely unacknowledged, that a pivotal turning point has been reached; several key technologies have achieved long-awaited maturity, most notably predictive immunoinformatic methods and post-genomic strategies.

In another popular definition, immunomics can stand as a synonym for system biology techniques applied to the study of Immunology. For many scientists, immunology is the pre-eminent example of systems behaviour in biology. Of course, the whole of biology – indeed the whole of the physical universe – behaves as a system, and exhibits characteristic systems behaviour. Since the immune system is innately hierarchical and exhibits confounding complexity at each tier of this cascading or branching hierarchy, as a system it can be said to exhibit emergent behaviour at all levels. Yet at the heart of the immune system are arrays of straightforward if not uncomplicated molecular recognition events, each of which is essentially indistinguishable from other biomolecular interactions. It is only when our frail and constrained mortal minds study something do we see the components in isolation. This is the glory – and the limitation – of reductionist approaches to scientific insight and discovery.

Since the discovery of antibodies and MHC restriction, humoural and cellular immunologists have sought to understand the nature of these biomacromolecular interactions, seeking to analyse them in the most fundamental way. Systems biology seeks to analyse higher levels of the immune system with the same degree of rigour, by both analysing the system as it exhibits itself at these individual levels and by integrating detailed, low-level, small-scale molecular or mesoscopic information and more overtly macroscopic measurements with more intrinsically qualitative anatomical, functional, and phenotypic data. Thus, Systems Biology – or, in this context, Systems Immunomics – can be said to function at various length scales from the atomic to the macroscopic.

Biological systems, of which immunological systems are an example, are seldom binary entities on the whole organism scale, any more than their cascading sub-systems – be they organ, tissue, or cellular – are binary entities at subsidiary levels. They operate stochastically, subject to random fluctuations and exhibit clear non-linear behaviour. Immunology only truly manifests itself at the level of the whole organism, but at every intermediate level down to that of the molecule, significant and often unexpected emergent behaviour within experimental systems is observed.

Many tools exist within systems biology. Some tools are based on capitalising on the latent power of simulation, be that simulations of abstract theoretical or mathematical models or molecular simulations of precise descriptions of molecular system. Other tools are analytical tools that can be used together to effect the synthesis of competing thesis and antithesis through the integration of measured data.

The simplest types of systems model include network maps, which reticulate pathway components producing complex cellular representations akin to circuit diagrams, and so-called logical models, which describe immunological process in terms of sets of relatively simple rules, framed as Boolean logic, such as if X AND Y but NOT Z then A = 1. There are many other more complex and mathematically demanding models available; these include correlation models and kinetic modelling.

Correlation models are familiar to anyone who has ever tried statistical approaches to the prediction problem. Multiple linear regression or Partial Least Squares or neural networks, or, indeed, any of a hundred other data mining techniques, can be used to identify commonalities of exchange or cooperation within or between the measured outputs of different signalling or regulatory pathways. Kinetic models, on the other hand, try to picture the spatio-temporal behaviour of each and every individual component within the system. They are the zenith and apotheosis of complexity with the currently available approaches within systems biology.

It is also possible to combine these different kinds of model. One can use a hybrid of a kinetic model and one based on Boolean logic. This is particularly useful when one wishes to fuse data of different granularity. It is possible, for example, to build a detailed kinetic model for part of a pathway and then to fill in the lacuna within the available data by modelling the rest using much simpler Boolean models.

The word "bioinformatics" has formed part of the scientific lingua franca since the early 1990s; yet a simple and straightforward, and comprehensive and inclusive definition remains strangely elusive. A particularly succinct epitome of the discipline is: "Bioinformatics is the application of informatics methods to biological macromolecules."

There are many reasons for this failure of orismology, partly arising because bioinformatics is in constant flux; undergoing relentless change, growth, and differentiation: you cannot easily name something that is never still enough to describe.

Bioinformatics has greatly expanded over the years, allowing for both new sub-disciplines to emerge within it and for bioinformatics to merge with other disciplines producing new and exciting hydrids. Sub-disciplines have tended to focus on areas of applications, such as neuroinformatics, transcriptomics, or proteomics, while hybrids have included text mining or statistical genetics. Immunoinformatics is another important sub-discipline. It deals specifically with the unique problems of the immune system. Like Bioinformatics, immunoinformatics complements, but never replaces, practical experimentation. It helps, and in a systematic way, researchers to answer the key questions in the still highly empirical world of immunology.

The scope and focus of bioinformatics is constantly developing and expanding to encompass more and more new areas of application. However, it is clear that Bioinformatics concerns itself with medical, genomic, and biological information and supports both basic and clinical research. Bioinformatics develops computer databases and algorithms for accelerating, simplifying, and thus enhancing, research in bioscience. Within this, however, the nature and variety of different bioinformatic activities are hard to quantify. Bioinformatics is as much a melting pot of interdisciplinary techniques as it is a branch of information science: it operates at the level of protein and nucleic acid sequences, their structures, and their functions, using data from microarray experiments, traditional biochemistry, as well as theoretical biophysics.

Databases are a key component of research in bioinformatics and immunoinformatics. They have been so and will remain vital for the foreseeable future. They are, or should be, as much tools as the algorithms used to search, analyse, and interrogate them. Bioinformatics is largely concerned with data handling, mainly through the annotation of macromolecular sequence and structure databases. A number of chapters in this book describe the application and development of databases within the immunoinformatic domain. Chapters "IPD – The Immuno Polymorphism Database" and "The IMGT/HLA Database", by Professor Marsh and co-workers, describe two world-leading resources: IPD and IMGT/HLA. Chapter "Ontology Development for the Immune Epitope Database" by Bjorn Peters and colleagues neatly summarises on-going development of the IEDB database. Chapter "Databases and Web-Based Tools for Innate Immunity" extends and completes this strand by describing a variety of databases aimed at the archiving of data relating to the innate immune system.

The growth of Bioinformatics is a clear success story of the informatic applications in bioscience. The services of bioinformaticians remain much in demand by forward-thinking biologists of many kinds. As new genomes are, for example, sequenced, biologists and immunologists wish to know many things: where and what post-translational modification there are; the location of protein within a cell; which proteins will substrates for proteases or kinases or other enzymes; even down to the pKa of a particular residue within a certain enzyme active site residue. The list seems, and is, endless, or nearly so. Attempting to address all of these possibilities in a systematic and effective manner using experiment only would be prohibitive to the point of intractability, in terms of time, resource, and that most precious quantity of all: human labour. The only practical and practicable solution is the deployment of bioinformatics.

Bioinformatics focuses on analysing molecular sequence and structure data, molecular phylogenies, and the analysis of post-genomic data generated by genomics, transcriptomics, and proteomics. Bioinformatics seeks to find solutions to two key challenges. First, the prediction of Function from Sequence, which can be performed using global homology searches, motif databases searches, and the formation of multiple sequence alignments. Chapter "Discovery of Conserved Epitopes Through Sequence Variability Analyses" indicates the wisdom of this assertion; it addresses the prediction of conserved epitopes within an immunomics and immunoinformatic context. Chapter "Defining the Elusive Molecular Self" picks up on this with its analysis of the molecular nature of the immune self.

Secondly, the prediction of Structure from Sequence, which may be attempted using secondary structure prediction, threading, and comparative, or so-called homology, modelling. Chapter "Structural Immunoinformatics: Understanding MHC-Peptide-TR Binding" provides a lucent and definitive description of the use of 3-dimensional structural data, as derived from experiment and computation, within the province of immunoinformatic investigation. As yet, the full power of 3-dimensional data has not been realised. Structure-based computation based on dynamic simulation and hypothesis-guided modelling has so much to reveal, but as yet the potential is not matched by available computing resources. The next few years should see this approach beginning to bear fruit as more and more studies are undertaken.

It is also an implicit assumption that knowledge of a structure facilitates prediction of function. In reality, all predictions of function rely on identifying similarity between sequences or between structures. When this similarity is very high, and thus is intrinsically reliable, then useful inferences may be drawn, but as similarity falls away any conclusions that are inferred become increasingly uncertain and potentially misleading. Thus, provenance is everything; and provenance and annotation. Bioinformatics still concerns handling and analysing data, often basing the classification of sequences or structures into coherent groups on the rigorous annotation of macromolecular sequence and structure databases. The Tepidas system described in chapter "TEPIDAS: A DAS Server for Integrating T-Cell Epitope Annotations" addresses the integration of data sources for the rigorous and reliable annotation of T cell epitopes.

Vaccines were for so long a moribund market, yet they have recently re-emerged as the most hopeful growth area for the Pharmaceutical Industry. Public health requirements safeguard vaccine supply of vaccines and in the absence of competition – Influenza apart, only two to three manufacturers target each vaccine-preventable disease – this has led to a recent increase in unit price for specialty vaccines. The launch of pioneering products – Wyeths Prevnar or Merck's Gardasil, for example – together with a much more favourable regulatory framework has made vaccines a key focus of the biotechnology and pharmaceutical industries. Paediatric vaccines currently hold sway over the global market for vaccines, yet adult vaccines will help drive future growth. The cancer vaccine market, led by vaccines targeting cervical cancer, is the most lucrative area of vaccine development: by 2012, cancer vaccines will account for around 30% of all vaccine revenues. As discussed in chapter "Computational Vaccinology", Immunomics and Systems Immunomics, at least in their informatic and computational guise, have much to offer vaccine design and discovery and the still emergent science of Vaccinology.

Returning to our first theme, the term vaccinology is said by many to have been coined by Jonas Salk to distinguish the systematic scientific study of vaccines – and thus how to develop and discover them – from the practice of vaccination as a medical art. In recent times, another term, immunovaccinology has been adopted by some to further differentiate the study of vaccine discovery and development based on a sound understanding of immunology, if such a thing exists, from what many might consider the highly empirical, microbiology-based science of vaccinology, as practiced in year gone by. Davies and Flower give a concise examination of how immunoinformatics has and can impact upon the pursuance of a rational yet systematic approach to vaccine discovery.

The next stage in the evolution of Immunomics and Systems Immunomics will come as closer collaborative links are forged between immunoinformaticians and experimentalists searching for new and deeper understanding of immunology, within and between both academic and commercial organisations. Immunomics must be both client and provider acting as a consumer of existing techniques and as an inspiration for other techniques. In this regard, the prime acolyte is that branch of computer science known as artifical immune systems or AIS research. The power and potential of AIS is amply demonstrated by chapter "Tunable Detectors for Artificial Immune Systems: From Model to Algorithm". This looks at how the immune system is able to provide a metaphor in the development of tunable detectors.

Despite the need for more accurate prediction algorithms, able to cover ever more MHC alleles in ever more species, the lack of persuasive evaluations of known methods continues to hamper and stymie uptake of this technology. In order that Immunoinformatic approaches might one day become universally used by experimental immunologists, methods should be tested over a wide range of alleles, species, and sequence-distinct peptides, with their accuracy reaching a high statistical significance. This will be greatly facilitated by adoption of a cyclically and progressive process of using and refining models and experiments.

The effective implementation of immunoinformatic strategies within Immunomics and Systems Immunomics will deliver an unprecedented dividend of great if unquantifiable magnitude. Methods that accurately predict individual components of the immune response or allow us to model the behaviour of the whole system or part thereof will be the most vital of tools for tomorrow's immunologists and vaccinologist. Immunoinformatic prediction, within the broader system immunomics context, remains a grand scientific problem, being both challenging, and thus exciting, and of true practical value. Moreover, the proper realisation of Systems Immunomics requires not only a deep appreciation of immunological mechanisms but also requires one to integrate many other disciplines, both experimental and theoretical. To enable this requires more than improved methods and software, it necessitates building immunoinformatics into the basic strategy of immunological investigation and it needs the confidence of experimentalists to commit laboratory work on this basis. Within the context of immunomics and systems immunomics, the synergy of experimental and informatics-based disciplines will enhance significantly our ability to understand and manipulate immunology process, leading to the augmented discovery of new laboratory reagents and diagnostics, in addition to new biomarkers and candidate vaccines.

M.N. Davies
S. Ranganathan
D.R. Flower

Computational Vaccinology

Matthew N. Davies and Darren R. Flower

Introduction

For vaccines, it is the best of times and the worst of times. Mass vaccination and public sanitation are the two most effective prophylactic treatments forestalling the depredations of infectious disease. The greatest part of the vaccine story is that of smallpox. At its height, Smallpox killed 10% of Swedish children within their first year. In London, more than 3,000 died in a single smallpox epidemic in 1746, and during the period of 1760–1770, the city lost another 4% of its population to the disease. Even as recently as the late 1960s, there were 10–12 million cases in 31 countries, with two million deaths annually. Yet the disease is, apart from a few hopefully well-guarded stockpiles, a thing of the past. There have been no cases for most of last 30 years. It has been completely eradicated. Poliomyelitis or Polio has also been the target for a worldwide campaign designed to eradicate the disease by the year 2000. A program undertaken by the Pan American Health Organization eliminated polio in the Western Hemisphere in 1991. The Global Polio Eradication Program has dramatically reduced poliovirus transmission throughout the world. In 2003, only 784 confirmed cases of polio were reported globally. Today Polio remains endemic in only four countries: Nigeria, Afghanistan, Pakistan, and India.

Yet millions still die from disease preventable through vaccination. Infectious diseases are to blame for around 25% of global mortality, particularly in children under five. Annual figures are stark: pertussis accounts for 294,000 (aged under 5); tetanus for 198,000 (under 5) and 15,000 (aged over 5); Hib for 386,000 (under 5); Hepatitis B for 600,000 (over 5); Yellow Fever for 15,000 (under 5) and 15,000 (over 5); and diphtheria for 4,000 (under 5). Projections for as-yet-not-universally available vaccines estimate a consequently greater saving of human life: 449,000 for rotavirus vaccine and 1,612,000 for Pneumococcus vaccine. For example, influenza with an annual global estimate of half a million deaths. However, perhaps the most lamentable situation is Measles, which accounts for 540,000 (under 5) and 70,000 (over 5).

M.N. Davies (✉)
The Jenner Institute, University of Oxford, High Street, Compton, Berkshire, RG20 7NN, UK
e-mail: m.davies@mail.cryst.bbk.ac.uk

D.R. Flower et al. (eds.), *Bioinformatics for Immunomics*, Immunomics Reviews 3,
DOI 10.1007/978-1-4419-0540-6_1,

Measles is an acute viral disease. The Arabic physician Rhazes referred to it, saying that it was 'more dreaded than smallpox'. The eldest son of the famous late Victorian and Edwardian author, H Rider Haggard, who, among 68 novels, wrote *Allan Quatermain* and *She*, died from the disease aged 10. Oliva Dahl, daughter of Roald, died aged 7 from measles-induced encephalitis.

The leading annual causes of death are 2.9 millions for tuberculosis; 2.5 million for diarrhoeal illnesses, especially rotaviruses; a rapidly escalating 2.3 million for HIV/AIDS; and 1.08 millions of deaths for malaria. There are no effective vaccines for HIV (Girard et al. 2006) and Malaria (Vekemans and Ballou 2008), two of the WHO's big three global killers; indeed, there is little hope that such vaccines will appear in the foreseeable future. And the only vaccine licensed for the third major world disease, tuberculosis, is of limited efficacy (de Lisle et al. 2005). Add to this the 35 new, previously unknown infectious diseases identified in the past 25 years: HIV, Marburg's disease, SARS, Dengue, West Nile, and over 190 human infections with potentially pandemic H5N1 influenza.

Many infections caused by viruses have proved stubborn and recalcitrant threats to human health and wellbeing. 350 million people carry hepatitis B (HBV). 170 million carry Hepatitis C (HCV), and 40 million carry human immunodeficiency virus type 1 (HIV-1). Each year, 5–15% of the world's population become infected by a new variant of the influenza virus, causing 250,000–500,000 deaths. Latent Bacterial infection can be even higher: there are, for example, over two billion people infected with TB. It is a commonly held conception that new infectious diseases will emerge continually throughout the twenty-first century. We are threatened by parasitic diseases such as malaria, visceral leishmaniasis, tuberculosis, and emerging zoonotic infections, such as H5N1; antibiotic-resistant bacteria; and bioterrorism; a threat compounded by a growing world population, overcrowded cities, increased travel, climate change, and intensive food production.

Thus it is the best of times for vaccines because of their uncompromising success and the worst of times because so much remains to be done. There are now more than 30 licensed individual vaccines targeted against 26 infectious diseases, most of which are viral or bacterial in nature. About half of these 30 vaccines are in common use, and are, in the main, employed to prevent childhood infections. In the First World, annual mortality for diseases such as polio, diphtheria, or measles is less than 0.1%; and the lasting effects of vaccination work to greatly reduce the morbidity and mortality of disease, often conferring lifetime protection. Most vaccines target childhood infections or are used by travellers to tropical or sub-tropical regions; a significant minority combat disease in the developing world.

Within the pharmaceutical industry and academia, vaccines are also seeing the best and worst of times. Persistent infection, which includes HIV, Hepatitis B, hepatitis C, and TB, occurs when a pathogen evades or subverts T cell responses, is a key therapeutic target. At the other extreme are benign yet economically important infections, such as the common cold. Respiratory track infections remain the major cause of community morbidity and hospitalisation in the developed world: about 60% of GP referrals and cause the loss of a huge number of working days. Sporadic or epidemic respiratory infections are caused by over 200 distinct viruses, including coronaviruses, rhinoviruses, respiratory syncytial virus (better known as RSV), parainfluenza virus, influenza A and B, and cytomegalovirus. Anti-Allergy vaccination also offers great potential for successful commercial exploitation. This often relies on allergen-specific immunotherapy (STI) (Palma-Carlos et al. 2006), where a patient is administered increasing quantities of allergen to augment their natural tolerance. STI, though often effective, is very time consuming and is not universally applicable. Recombinant hypo-allergenic allergens are also of interest, as they can target specific immune cells. New agents for the prophylaxis and treatment of allergic disease are legion: recombinant proteins, peptides, immunomodulatory therapy, and DNA vaccines, which are particularly promising tools. Several anti-allergy DNA vaccines are being developed: including optimised expression of allergen genes, CpG-enrichment of delivery vectors, and the targeting of hypoallergenic DNA vaccines. Vaccines against the common cold or anti-allergy vaccines lie close to so-called life-style vaccines. None of these vaccines necessarily saves lives but does reduce hugely important economic effects of disease morbidity. Life-style vaccines target dental caries and drug addiction, as well as genetic and civilisation diseases, such as obesity.

Vaccination is also being used to tackle cancer. Gardasil, the new human papillomavirus vaccine (Hung et al. 2008), was licensed in 2006 with the goal of saving 4,000 deaths a year from cervical cancer. Cancer is the second greatest cause of death in the developed world after cardiovascular disease; yet most of the 250,000 deaths from cervical cancer occur in the Third World. Cancer treatment typically involves a combination of chemotherapy, radiotherapy, and surgery. While treating primary tumours this way is largely successful, preventing the metastatic spread of disease is not. Cancer vaccines are attractive, both clinically and commercially, since they exploit immunity's ability to recognise and destroy tumours. Tumour cells express several mutated or differentially expressed antigens, enabling the immune system to discriminate between non-malignant and cancerous cells. Tumour antigens form the basis of both subunit and epitope-based vaccines. Host immune system responses to tumour-antigen cancer vaccines are often weak, necessitating the use of adjuvants.

Thus we can see that, biomedically speaking at least, vaccines are indeed having the best and worst of times: best because of the enormous opportunity for them to treat an ever-widening tranche of diseases and worst because of the inadequacies of existing techniques used to foster vaccine development. Vaccinology has, until relatively recently, been a primarily empirical science, relying on tried-and-tested – yet poorly understood – approaches to vaccine development. As a consequence of this, few effective vaccines were developed and deployed during the 150 years following Jenner: most targets remained inaccessible to science to the emerging science of vaccinology. The success of a vaccine can be measured by its strength, its specificity, the duration of the immune response, and its capacity to create immunological memory. A vaccine is a molecular or supramolecular agent which can elicit specific protective immunity and ultimately mitigate the effect of subsequent infection. Vaccination is the use of a vaccine, in whatever form, to produce active prophylactic immunity in a host organism.

Vaccines have taken many forms. Until recently, they have been attenuated or inactivated whole pathogen vaccines such as anti-tuberculosis BCG or Sabin's vaccine against Polio. Safety difficulties have led to the subsequent development of other strategies for vaccine development. The most successful alternative has focused on the antigen – or subunit – vaccine, such as recombinant Hepatitis B vaccine (Ebo et al. 2008). Vaccines based around sets of epitopes have also gained ground in recent years. They can be delivered into the host in many ways: as naked DNA vaccines, using live viral or bacterial vectors, and via antigen-loaded antigen presenting cells. Adjuvants are substances, such as alum, which are used with weak vaccines to increase immune responses (O'Hagan et al. 2001).

With more and more pathogen genomes being fully or partially determined, it has become imperative to develop reliable *in silico* methods able to identify potential vaccine candidates within microbial genomes. While it is possible to assess in the laboratory those properties of vaccine which make it successfully, it is not practical to do on the scale of a large pathogen genome. Immunomics is a solution to this dilemma. It is a post-genomic systems biology approach to immunology that explores mechanistic aspects of the immune system (De Groot 2006). It subsumes immunoinformatics and computational vaccinology, combining several fields, including genomics, proteomics, immunology and clinical medicine. To date, a key focus of immunomics has been the development of algorithms for the design and discovery of new vaccines. Here we outline currently available techniques and software for vaccine discovery as well as examples of how such algorithms can be applied. We concentrate on four areas: antigen prediction, epitope prediction, vector design, and adjuvant identification.

Epitope Prediction

Complex microbial pathogens, such as Mycobacterium tuberculosis, can interact within the immune system in a multitude of ways (McMurry et al. 2005). For a vaccine to be effective it must invoke a strong response from both T Cells and B Cells; therefore, epitope mapping is a central issue in their design. *In silico* prediction methods can accelerate epitope discovery greatly. B Cell and T Cell epitope mapping has led to the predictive scanning of pathogen genomes for potential epitopes (Pizza et al. 2000a). There are over 4,000 proteins in the TB genome; this means that experimental analysis of host-pathogen interactions would be prohibitive in terms of time, labour, and expense.

T Cell Epitope Prediction

T cell epitopes are antigenic peptide fragments derived from a pathogen that, when bound to a Major Histocompatibility Complex (MHC) molecule, interact with T Cell receptors after transport to the surface of an Antigen-Presenting Cell. If sufficient quantities of the epitope are presented, the T Cell may trigger an adaptive immune response specific for the pathogen. MHC Class I and Class II molecules form complexes with different types of peptide. The Class I molecule binds a peptide of 8–15 amino acids in length within a single closed groove. The peptide is secured largely through interactions with anchoring residues at the N- and C-termini of the peptide, while the central region is more flexible (Rammensee et al. 1999). Class II peptides vary in length from 12 to 25 amino acids and are bound by the protrusion of peptide side chains into cavities within the groove and through a series of hydrogen bonds formed between the main chain peptide atoms and the side chains atoms of the MHC molecule (Jardetzky et al. 1996). Unlike the Class I molecule, where the binding site is closed at either end, the peptide can extend out of both open ends of the binding groove.

Experimentally determined IC_{50} and BL_{50} affinity data have been used to develop a variety of MHC-binding prediction algorithms, which can distinguish binders from non-binders based on the peptide sequence. These include motif-based systems, Support Vector Machines (SVMs) (Donnes and Elofsson 2002; Liu et al. 2006; Wan et al. 2006), Hidden Markov Models (HMMs) (Noguchi et al. 2002), QSAR analysis (Doytchinova et al. 2005), and structure-based approaches (Davies et al. 2006; Davies et al. 2003; Wan et al. 2004). MHC-binding motifs are a straightforward and easily comprehended method of epitope detection, yet produce many false-positive and many false-negative results. Support Vector Machines (SVMs) are machine learning algorithms based on statistical theory that seeks to separate data into two distinct classes (in this case binders and non-binders). HMMs are statistical models where the system being modelled is assumed to be a Markov process with unknown parameters. In an HMM, the internal state is not visible directly, but variables influenced by the state are. HMMs aim to determine the hidden parameters from observable ones. An HMM profile can be used to determine those sequences with 'binder-like' qualities. QSAR analysis techniques have been used to refine the peptide interactions with the MHC Class I groove by incrementally improving and optimising the individual residue-to-residue interactions within the binding groove. This has led to the design of so-called superbinders that minimise the entropic disruption in the groove and are therefore able to stabilise even disfavoured residues within so-called anchor positions. Finally, molecular dynamics has been used to quantify the energetic interactions between the MHC molecule and peptide for both Class I and Class II by analysis of the three-dimensional structure of the MHC-peptide complex.

Many programs that are able to facilitate the design of optimised vaccines are now available. In this section, some of the most effective algorithms for each form of vaccine design are discussed. For T Cell epitope prediction, many programs are available. A sensible approach for a new user would be to use MHCBench (Salomon and Flower 2006), an interface developed specifically for evaluating the various MHC-binding peptide prediction algorithms. MHCBench allows users to compare the performance of various programs with both threshold-dependent and -independent parameters. The server can also be extended to include new methods for different MHC alleles.

B Cell Epitope Prediction

B cells generate antibodies when stimulated by helper T cells as part of the adaptive immune response. The antibodies act to bind and neutralise pathogenic material from a virus or bacterium. Individual antibodies are composed of two sets of heavy and light chains. Each B cell produces a unique antibody due to the effects of somatic hypermutation and gene segment rearrangement. Those cells, within the primary repertoire whose antibodies convey antigen recognition, are selected for clonal expansion, an iterative process of directed hypermutation and antigen-mediated selection. This facilitates the rapid maturation of antigen-specific antibodies with a high affinity for a specific epitope. A B cell appropriate to deal with a specific infection is selected and cloned to deal with the primary infection, and a population of the B cell is then maintained in the body to combat secondary infection. It is the capacity to produce a huge variety of different antibodies that allows the immune system to deal with a broad range of infections.

B Cell prediction is more problematic due to the difficulties in correctly defining both linear and discontinuous epitopes from the rest of the protein. The epitope of a B Cell is defined by the discrete surface region of an antigenic protein bound by the variable domain of an antibody. The production of specific antibodies for an infection can boost host immunity in the case of both intracellular and extracellular pathogens. The antibody's binding region is composed of three hypervariable loops that can vary in both length and sequence so that the antibodies generated by an individual cell present a unique interface (Blythe and Flower 2004). All antibodies contain two antigen-binding sites, composed of complementary determining (CDR) loops. The three CDR loops of the heavy and light chains form the 'paratope', the protein surface which binds to the antigen. The molecular surface that makes specific contact with the residues of the paratope is termed an 'epitope'. A B cell epitope can be an entire molecule or a region of a larger structure. The study of the paratope–epitope interaction is a crucial part of immunochemistry, a branch of chemistry that involves the study of the reactions and components on the immune system.

Despite the extreme variability of the region, the antibody-binding site is more hydrophobic than most protein surfaces with a significant predilection for tyrosine residues. B Cell epitopes can be divided into continuous (linear) and discontinuous (conformational), the latter being regions of the antigen separated within the sequence but brought together in the folded protein to form a three-dimensional interface. Another problem with B cell epitopes relates to the fact that they are commonly divided into two groups: continuous epitopes and discontinuous epitopes. Continuous epitopes correspond to short peptide fragments of a few amino acid residues that can be shown to cross-react with antibodies raised against the intact protein. Since the residues involved in antibody binding represent a continuous segment of the primary sequence of the protein they are also referred to as 'linear' or 'sequential' epitopes. Studies have shown that this class of epitope often contains residues that are not implicated in antibody interaction, while some residues play a more important role than others in antibody binding. Discontinuous epitopes are composed of amino acid residues that are not sequential in the primary sequence of a protein antigen but brought into spatial proximity by the three-dimensional folding of the peptide chain (Greenbaum et al. 2007).

There is considerable interest in developing reliable methods for predicting B cell epitopes. However, to date, the amino acid distribution of the complementary antigen surface has been difficult to characterise, presenting no unique sequential or structural features upon which to base a predictive system. It is partly for this reason that B Cell epitope has lagged far behind T Cell prediction in terms of accuracy but also because most of the data upon which predictions are based remain open to question due to the poorly understood recognition properties of cross-reactive antibodies. One of the central problems with B cell epitope prediction is that the epitopes themselves are entirely context dependent. The surface of a protein is, by definition, a continuous landscape of potential epitopes that is without borders. Therefore both epitope and paratope are fuzzy recognition sites, forming not a single arrangement of specific amino acids but a series of alternative conformations. In this instance, a binary classification of binder and non-binder may simply not reflect the nature of the interaction. A factor also to be considered is that the average paratope consists of only a third of the residues within the CDR loops, suggesting the remaining two-thirds could potentially bind to an antigen with an entirely different protein surface.

Often a short length of amino acids can be classified as a continuous epitope, though in fact it may be a component of a larger discontinuous epitope; this can be a result of the peptide representing a sufficient proportion of the discontinuous epitope to enable cross-reaction with the antibody. Since the majority of antibodies raised against complete proteins do not cross-react with peptide fragments derived from the same protein it is thought that the majority of epitopes are discontinuous. It is estimated that approximately 10% of epitopes on a globular protein antigen are truly continuous in nature. In spite of the this, the majority of research into B cell epitope prediction has focussed largely on linear peptides on the grounds that they are discrete sequences and easier to analyse. This can only be resolved by examination of the three-dimensional structure of the protein where the distinction between the continuous and discontinuous forms is not relevant.

Initial research into B cell epitope prediction looked for common patterns of binding or 'motifs' that characterise epitope from non-epitopes. Unfortunately, the wide variety of different epitope surfaces that can be bound made it impossible to determine any such motifs. More sophisticated machine learning approaches such as Artificial Neural Networks have also been applied but never with an accuracy exceeding 60%. More recently, structural analysis of known antigens has been used to determine the surface accessibility of residues as a measure of the probability that they are part of an epitope site. Despite these fundamental limitations, several B Cell epitope prediction programs are available including Discotope (Andersen et al. 2006), 3DEX (Schreiber et al. 2005) and CEP (Kulkarni-Kale et al. 2005). Both CEP and Discotope measure the surface accessibility of residues although neither has been developed to the point where they can identify coherent epitope regions rather than individual residues. A recent review of B cell epitope software

(Ponomarenko and Bourne 2007) calculated the A_{ROC} curves for the evaluated methods were about 0.6 (indicating 60% accuracy) for DiscoTope, ConSurf (which identifies functional regions in proteins), and PPI-PRED (protein–protein interface analysis) methods, while protein–protein docking methods were in the region of 65% accuracy, never exceeding 70%. The remaining prediction methods assessed were all close to random. In spite of this, the increasing number of available antigen–antibody structures combined with sophisticated techniques for structural analysis suggests a more methodical approach to the study interface will yield a better understanding of what surfaces can and cannot form stable epitopes. The proposed research will take several different approaches to this problem that will lead to a more comprehensive understanding of antibody–antigen interactions.

Computational Identification of Virulence Factors

The word antigen has a wide meaning in immunology. We use it here to mean a protein, specifically one from a pathogenic micro-organism, which evokes a measurable immune response. Pathogenic proteins in bacterial are often acquired, through a process summarised by the epithet horizontal transfer, in groups. Such groups are known as pathogenicity islands. The unusual G+C content of genes and particularly large gene clusters is tantamount to a signature characteristic of genes acquired by horizontal transfer. Genome analyses at the nucleic acid level can thus allow the discovery of pathogenicity islands and the virulence genes they encode.

Perhaps the most obvious antigens are virulence factors (VF): proteins which enable a pathogen to colonise a host or induce disease. Analysis of pathogens – such as *Vibrio cholerae* or *Streptococcus pyogenes* – has identified coordinated 'systems' of toxins and virulence factors which may comprise over 40 distinct proteins. Traditionally, VFs have been classified as adherence/colonisation factors, invasions, exotoxins, Transporters, iron-binding Siderophores, and miscellaneous cell surface factors. A broader definition, groups VFs into three: 'true' VF genes, VFs associated with the expression of 'true' VF genes, and VF 'life-style' genes required for colonisation of the host (Guzmán et al. 2008).

Several databases exist which archive VFs. The Virulence Factors Database (VFDB) contains 16 characterised bacterial genomes with an emphasis on functional and structural biology and can be searched using text, BLAST, or functional queries (Yang et al. 2008). The ClinMalDB-US database is being establishing following the discovery of multi-gene families encoding VFs within the subtelomeric regions of *P. falciparum* (Mok et al. 2007) and *P. vivax* (Merino et al. 2006). TVFac (Los Alamos National Laboratory Toxin & Virulence Factor database) contains genetic information on over 250 organisms and separate records for thousands of virulence genes and associated factors. The Fish Pathogen Database, set up by the Bacteriology & Fish Diseases Laboratory, has identified over 500 virulence genes using fish as a model system. Pathogens studied include *Aeromonas hydrophila*, *Edwardsiella tarda*, and many *Vibrio* species.

Candida albicans Virulence Factor (CandiVF) is a small species-specific database that contains VFs which may be searched using BLAST or a HLA-DR Hotspot Prediction server (Tongchusak et al. 2008). PHI-BASE is a noteworthy development, since it seeks to integrate a wide range of VFs from a variety of pathogens of plants and animals (Winnenburg et al. 2008). Obviously, antigens need not be virulence factors and another nascent database is intending to capture a wider tranche of data. We are helping to develop the AntigenDB database [http://www.imtech.res.in/raghava/antigendb/] which will aid considerably this endeavour.

Identifying Antigens *In Silico* Using Subcellular Location Prediction

Historically, antigens have been supposed to be secreted or exposed membrane proteins accessible to surveillance of the immune system. Subcellular location prediction is thus a key approach to predicting antigens. There are two basic kinds of prediction method: manual construction of rules of what determines subcellular location and the application of data-driven machine learning methods, which determine factors that discriminate between proteins from different known locations. Accuracy differs markedly between different methods and different compartments, mostly due to a paucity of data. Data used to discriminate between compartments include: the amino acid composition of the whole protein; sequence derived features of the protein, such as hydrophobic regions; the presence of certain specific motifs; or a combination thereof.

Different organisms evince different locations. PSORT is a knowledge-based, multi-category prediction method, composed of several programs, for subcellular location (Rey et al. 2005); it is often regarded as a gold standard. PSORT I predicts 17 different subcellular compartments and was trained on 295 different proteins, while PSORT II predicts ten locations and was trained on 1,080 yeast proteins. Using a test set of 940 plant proteins and 2,738 non-plant proteins, the accuracy of PSORT I and II was 69.8% and 83.2%, respectively. There are several specialised versions of PSORT. iPSORT deals specifically with secreted, mitochondrial and chloroplast locations; its accuracy is 83.4% for plants and 88.5% for non-plant. PSORT-B only predicts bacterial subcellular locations. It reports precision values of 96.5% and recall values of 74.8%. PSORT-B is a multi-category method which combines six algorithms using a Bayesian Network.

Among binary approaches, arguably the best method is SignalP, which employs neural networks and predicts N-terminal Spase-I-cleaved secretion signal sequences and their cleavage site (Emanuelsson et al. 2007). The signal predicted is the type-II signal peptide common to both eukaryotic and prokaryotic organisms, for which there is wealth of data, in terms of both quality and quantity. A recent enhancement of SignalP is a Hidden Markov Model version able to discriminate uncleaved signal anchors from cleaved signal peptides.

One of the limitations of SignalP is over-prediction, as it is unable to discriminate between several very similar signal sequences, regularly predicting membrane proteins and lipoproteins as type-II signals. Many other kinds of signal sequence exist. A number of methods have been developed to predict lipoproteins, for example. The prediction of proteins that are translocated via the TAT-dependent pathway is also important but is not addressed yet in any depth.

The Many Successes of Reverse Vaccinology

Reverse vaccinology is a principal means of identifying subunit vaccines and involves a considerable computational contribution. Conventional experimental approaches cultivate pathogens under laboratory conditions, dissecting them into their components, with proteins displaying protective immunity identified as antigens. However, it is not always possible to cultivate a particular pathogen in the lab nor are all proteins expressed during infection are easily expressed *in vitro*, meaning that candidate vaccines can be missed. Reverse vaccinology, by contrast, analyses a pathogen genomes to identify potential antigens and is typically more effective for prokaryotic than eukaryotic organisms.

Initially, an algorithm capable of identifying Open Reading Frames (ORFs) scans the pathogenic genome. Programs that can do this include ORF-FINDER (Rombel et al. 2003), Glimmer (Delcher et al. 1999), and GS-finder (Ou et al. 2004). Once all ORFs have been identified, proteins with the characteristics of secreted or surface molecules must be identified. Unlike the relatively straightforward task of identifying ORFs, selecting proteins liable to immune system surveillance is challenging. Programs such as ProDom (Servant et al. 2002), Pfam (Bateman et al. 2000), and PROSITE (Falquet et al. 2002) can identify sequence motifs characteristic of certain protein families and can thus help predict if a protein belongs to an extracellular family of proteins.

We have developed VaxiJen [http://www.jenner.ac.uk/VaxiJen/] that implements a statistical model able to discriminate between candidate vaccines and non-antigens, using an alignment-free representation of the protein sequence (Doytchinova and Flower 2007). Rather than concentrate on epitope and non-epitope regions, the method used bacterial, viral, and tumour protein datasets to derive statistical models for predicting whole protein antigenicity. The models showed prediction accuracy up to 89%, indicating a far higher degree of accuracy than has, for example, been obtained previously for B Cell epitope prediction. Such a method is an imperfect beginning; future research will yield significantly more insight as the number of known protective antigens increases.

The NERVE program has been developed to further automate and refine the process of reverse vaccinology, in particular the process of identifying surface proteins. In NERVE, the processing of potential ORFs is a six-step process. It begins with the prediction of subcellular localisation, followed by the calculation of probability of the protein being adhesion, the identification of TM domains, a comparison with the human proteome and then with that of the selected pathogen, after which the protein is assigned a putative function. The vaccine candidates are then filtered and ranked based upon these calculations. While it is generally accepted that determining ORFs is a relatively straightforward process, the algorithm used to define extracellular proteins from other proteins needs to be carefully selected. One of the most effective programs that can be used for this purpose is HensBC, a recursive algorithm for predicting the subcellular location of proteins. The program constructs a hierarchical ensemble of classifiers by applying a series of if-then rules. HensBC is able to assign proteins to one of four different types (cytoplasmic, mitochondrial, nuclear, or extracellular) with approximately 80% accuracy for Gram-negative bacterial proteins. The algorithm is non-specialised and can be applied to any genome. Any protein identified as being extracellular could be a potential vaccine candidate.

The technique of reverse vaccinology was pioneered by a group investigating *Neisseria meningitides*, the pathogen responsible for sepsis and Meningococcal meningitis. Vaccines based upon the capsular proteins have been developed for all the serotypes with the exception of subgroup B. The *Neisseria meningitides* genome was scanned for potential ORFs (Tettelin et al. 2000; Pizza et al. 2000b). Out of the 570 proteins that were identified, 350 could be successfully expressed *in vitro* and 85 of these were determined to be surface exposed. Seven identified proteins conferred immunity over a broad range of strains within the natural *N. meningitidis* population, demonstrating the viability of *in silico* analysis as an aid to finding candidates for the clinical development of a MenB vaccine. Another example of the successful application of reverse vaccinology is *Streptococcus pneumoniae*, a major cause of sepsis, pneumonia, meningitis, and otitis media in young children (Wizemann et al. 2001; Maione et al. 2005). Mining of the genome identified 130 potential ORFs with significant homology to other bacterial surface proteins and virulence factors. 108 of 130 ORFs were successfully expressed and purified; six proteins were found to induce protective antibodies against pneumococcal challenge in a mouse sepsis model. All six of these candidates showed a high degree of cross-reactivity against the majority of capsular antigens expressed in vivo and which are believed to be immunogenic in humans.

Another example is *Porphyromonas gingivalis* is a gram-negative anaerobic bacterium present in subgingival plaques present in chronic adult periodontitis, an inflammatory disease of the gums. Shotgun sequences of the genome identified approximately 370 ORFS (Ross et al. 2001). Seventy-four of these had significant global homology to known surface proteins or an association with virulence. Forty-six had significant similarity with other bacterial outer membrane proteins. Forty-nine proteins were identified as surface proteins using PSORT and 22 through motif analysis. This generated 120 unique proteins sequences, 40 of which were shown to be positive for at least one of the sera. These were used to vaccinate mice, with only two of the antigens demonstrating significant protection. *Chlamydia pneumoniae* is an obligate intracellular bacterium associated with respiratory infections, cardiovascular and atherosclerotic disease. 141 ORFS were selected through *in silico* analysis (Montigiani et al. 2002) and 53 putative surface-exposed proteins identified. If reverse vaccinology is applied appropriately in vaccine design, it can save enormous amounts of money, time, and wasted labour.

Developing Vectors for Vaccines Delivery

Safe and effective methods of gene delivery have been sought for 30 years. Viral delivery of genes has effectively targeted inter alia haemophilia, coronary heart disease, muscular dystrophy, arthritis, and cancer. Despite their immanent capacity to transfer genes into cells, concerns over safety, manufacturing, restricted targeting ability and plasmid size have limited deployment of effective and generic gene therapy approaches. This remains a key objective for vaccinology. Vectors for gene therapy and vaccines differ in their requirements, yet both must overcome issues of targeting, plasmid cargo, and adverse immunogenicity. For example, up to 10% of the vaccinia genome can be replaced by DNA coding for antigens from other pathogens. The resulting vector generates strong antibody and T cell responses, and is protective. Viruses commonly used as vectors include Poxviruses, Adeno, varicella, polio, and influenza. Bacterial vectors include both *Mycobacterium bovis* and Salmonella. Adding extra DNA coding for large molecule adjuvants greatly can exacerbate antibody or T cell responses.

Successful transfection is hampered by DNA degradation within and outside the cell, inadequate cell penetration, poor intracellular trafficking, and inefficient nuclear localisation. Gene delivery requires both vector escape from digestion in late endosomes and nuclear translocation. Caveolin-dependent endocytosis, phagocytosis, and macropinocytosis do not transfer of material to the endo-lysosomal pathway. Some internalised material is released into the cytosol, through unknown mechanisms. However creating vectors with such desirable properties is difficult and their effectiveness may be compromised by their capacity to down-regulate other immune responses. The efficient and rational design of effective vaccine vectors is an area where informatic techniques could play a large role.

Similar to, yet simpler than, viral vectors are so-called DNA vaccines; they are plasmids capable of expressing antigenic peptide within the host (Babiuk et al. 2000). They are an attractive alternative to conventional vaccines, generating both a cellular and a humoral immune response, which are effective versus intracellular pathogens. The efficiency of a DNA vaccine has been successfully enhanced using codon optimisation (Babiuk et al. 2003), CpG motif engineering (Uchijima et al. 1998), (Klinman et al. 1997), and the introduction of promoter sequences (Booth et al. 2007), (Lee et al. 1997). Codon optimisation has been most effective in enhancing protein expression efficiency. Codons optimal for Translation are those recognised by abundant tRNAs (Xu et al. 2001). Within a phylogenetic group, codon frequency is highly correlated with gene expression levels. Immunogenicity depends upon effective translation and transcription of the antigen; it is possible to enhance this by selecting optimal codons for the vaccine.

The most comprehensive approach to vaccine optimisation is taken by DyNAVacs, an integrative bioinformatics tool that optimises codons for heterologous expression of genes in bacteria, yeasts, and plants (Henry and Sharp 2007). The program is also capable of mapping restriction enzyme sites, primer design, and designing therapeutic genes. The program calculates the optimal code for each amino acid encoded by a stretch of DNA by using codon usage table, which contains codon frequencies for a variety of different genomes.

A similar technique, CpG optimisation, may be used to optimise the codons in respect to CG dinucleotides. Pattern recognition receptors that form part of the innate immune system can often distinguish prokaryotic DNA from eukaryotic DNAs by detecting unmethylated CpG dinucleotides in particular base contexts, which are termed 'CpG motifs'. The presence of such motifs in the sequence can be highly advantageous so long as it does not interfere with the process of codon optimisation.

Discovery of Adjuvants and Immunomodulators

Another technique for optimising the efficacy of vaccines is to develop an efficient adjuvant. Adjuvants are defined as any chemical which is able to enhance an immune response when applied simultaneously with a vaccine and thus improve the efficacy of vaccination (Harish et al. 2006); Singh and O'Hagan 2002. It is possible that some adjuvants act as immune potentiators, triggering an early innate immune response that enhances the vaccine effectiveness by increasing the vaccine uptake. Adjuvants may also enhance vaccination by improving the depot effect, the co-localisation of the antigen, and immune potentiators by delaying the spread of the antigen from the site of infection so that absorption occurs over a prolonged period (Stills 2005). Aluminium hydroxide or Alum is the only adjuvant currently licensed in humans. Aluminium-based adjuvants prolong antigen persistence due to the depot effect, as well as stimulating the production of IgG1 and IgE antibodies (Gupta 1998) and triggering the secretion of interleukin-4. There are also several small-molecule, drug-like adjuvants, such as imiquimod, resiquimod, and other imidazoquinolines (Singh and Srivastava 2003; Schijns 2003; Iellem et al. 2001). Other small molecules that have been investigated for adjuvant properties include Monophosphoryl-Lipid A, muramyl dipeptide, QS21, PLG, and Seppic ISA-51 (Schijns 2003). In many cases, the adjuvant molecules have displayed toxic properties or showed poor adsorption making them unsuitable for use. Thus there is a great demand for new compounds that can be used as adjuvants.

Chemokine receptors are a family of G-protein-coupled receptors (GPCRs) that transduce chemokines, leukocyte chemoattractant peptides that are secreted by several cell types in response to inflammatory stimuli (Charoenvit et al. 2004a; Hedrick and Zlotnik 1996; Luster 1998). GPCRs are a superfamily of transmembrane proteins responsible for the transduction of a variety of endogenous extracellular signals into an intracellular response (Locati and Murphy 1999; Christopoulos and Kenakin 2002; Gether et al. 2002). Activation of the chemokine receptors triggers an inflammatory response by inducing migration of the leukocytes from circulation to the site of injury or infection. The receptors play a pivotal role in angiogenesis, haematopoiesis, brain and heart development, and there is also evidence that CCR5 precipitates the entry of HIV-1 into CD4+ T cells by the binding of the viral envelope protein gp120 (Bissantz 2003; Deng et al. 1996). There are 18 chemokine receptors and over 45 known chemokine ligands. The chemokines can be divided into the CC and CXC family, the former contains two cysteine residues adjacent within the protein sequence while in the latter they are separated by a single amino acid. CCR4 is a chemokine receptor expressed on Th2-type CD4+ T cells and has been linked to allergic inflammation diseases such as asthma, atopic dermatitis, and allergic rhinitis. There are two chemokines which bind the CCR4 receptor exclusively, CCL22 and CCL17 (Feng et al. 1996). Inhibition of the two ligands has been shown to reduce the migration of T cells to sites of inflammation, suggesting than any CCR4 antagonist could provide an effective treatment for allergic reactions, specifically in the treatment of asthma. Both anti-CCL17 and anti-CCL22 antibodies have been observed to have efficacy, the property that enables a molecule to impart a pharmacological response, in murine asthma models.

It is possible for the CCR4 receptor to act as an adjuvant due to its expression by regulatory T cells (Tregs) that normally downregulate an immune response (Chvatchko et al. 2000). The Tregs inhibit dendritic cell maturation and thus down-regulate expression of the co-stimulatory molecule. A successful CCR4 antagonist would therefore be able to enhance human T cell proliferation in an *in vitro* immune response model by blocking the Treg proliferation. This suggests that an effective CCR4 antagonist would have the properties of an adjuvant. A combination of virtual screening and experimental validation has been used to identify several potential adjuvants capable of inhibiting the proliferation of Tregs. Small-molecule adjuvant discovery is amenable to techniques used routinely by the pharmaceutical industry. Three-dimensional virtual screening is a fast and effective way of identifying molecules by docking a succession of ligands into a defined binding site (Lieberam and Forster 1999). A large database of small molecules can be screened quickly and efficiently in this way. Using 'targeted' libraries containing a specific subset of molecules is often more effective. It is possible to use 'privileged fragments' to construct combinatorial libraries, those which are expected to have increased probability of success. A pharmacophore is a specific three-dimensional map of biological properties common to all active confirmations of a set of ligands exhibiting a particular activity that can be used to discover new molecules with similar properties. Several small molecules that have been investigated for adjuvant properties in this way (Schellhammer and Rarey 2004). More recently, molecules that selectively interfere with chemokine-mediated T Cell migration have shown the potential to act as adjuvants by down-regulating the expression of co-stimulatory molecules, limiting T Cell activation. Small-molecule chemokine receptor antagonists have been identified and shown to be effective at blocking chemokine function in vivo (Charoenvit et al. 2004b; Godessart 2005), although to date no compound has reached a phase II clinical trial.

Discussion

In 1900, prime causes of human mortality encompassed influenza, enteritis, diarrhoea, and pneumonia: together accounting for over 30% of fatalities. Conversely, cancer and heart disease were responsible for only 12%. Compare that to the final

25 years of the seventeenth century, when average life expectancy was below 40. Principal causes of death were again infectious disease: smallpox, tuberculosis, malaria, yellow fever, and dysentery, which affected adult and children alike. Seemingly little had changed in the intervening 150 years. Today, the picture is very different, at least in developed countries. Infectious disease accounts for below 2% of deaths. Chronic disease now account for over 60% of deaths in the First World.

To a first approximation, life expectancy has, on average, escalated consistently throughout the last few thousand years. Obviously, a few great epidemic diseases – principally the Black Death – have, on occasion, made not insignificant dents in this inexorable upward progression. Citizens of the Roman Empire enjoyed a mean life expectancy at birth of about 22 years. By the Middle Ages, this had, in Europe at least, increased to be about 33 years. By the middle of the nineteenth century general life expectancy had risen to roughly 43 years. In the early 1900s, mean life spans in more developed countries ranged from 35 to 55. Life expectancy has accelerated over the last hundred years or so and today over 40 countries have an average life expectancy exceeding 70 years. The average life span across the whole human population is somewhat lower, however; it is estimated at 64.8 years – 63.2 years for men and 66.47 years for women.

Iceland heads the most recent 2003 league table with a mean life expectancy 78.7 years, next is Japan (78.4 years), followed by Sweden (77.9 years), then Australia (77.7 years). Next come Israel and Switzerland jointly with 77.6 years, Canada (77.4 years), then Italy (76.9 years), New Zealand and Norway jointly (76.8 years), and then Singapore with a mean life expectancy of 76.7 years. The United Arab Emirates is in tenth place with 76.4 years, followed by Cyprus (76.1 years), and in joint twelfth place Austria and the United Kingdom (76.0 years). Perversely, perhaps, the richest and most economically successful country on Earth, The United States of America, only manages twentieth place (74.6); while the newly emergent tiger economies of China (69.9 years, 41st place) and India (61.8; 77th place), which threaten to dominate the economic and fiscal landscapes of the coming century, come even lower on the list.

The population in many developed countries is now said to be ageing, putting burgeoning pressure on social welfare systems. However, this so-called ageing is only comparative, and comparative to the past. In fact, around 27% of the global population are aged 14 or under; of this, 0.91 billion are male and 0.87 billion are female. Only an estimated 100 million children attend school. The gender imbalance also varies across the world. In Hennan province, the most populous region of China, the balance is skewed 118:100 in favour of boys, probably due to the elective abortion of female foetuses; the average in industrialised nations of the First World is 103–108:100. The global average is 104:100.

65.2% of the world's population are between the ages of 15 and 64, with a ratio of 2.15 billion men to 2.10 billion women. Worldwide, only 7.4% of people are actually 65 and over, with a balance of men to women of 0.21–0.27 billion. However, when compared to centuries past, this shift is remarkable; one might almost call it a lurch, it has been so rapid. One way in which this shift manifests itself is in the vastly increased longevity of the individual, as well as an increasing proportion of the population reaching old age. This phenomenon is, in part, a by product of the First World's ever more comfortable, ever more urbanised post-industrial environment. Coupled to decades of better nutrition, medical advances in both treatment regimes and medicines will allow an ever-increasing section of the population to exploit their own genetic predisposition to long life. In North America, it has been estimated that by the middle of the century, those living beyond the age 100 would number over 100,000. On this basis, some have predicted that the human life span will be routinely stretched to 120. However, the total global population of so-called super-centenarian – those living beyond 110 – is roughly 80; while only one in two billion will live beyond 116.

With this growth in life expectancy has come a concomitant growth in the diseases of old age. These include hitherto rare, or poorly understood, neurodegenerative diseases, such as Parkinson's or Alzheimer's disease which proportionally affect the old more, cardiovascular diseases, and stroke, the prevalence of which is also increasing.

Patterns of disease have changed over the past hundred years and will change again in the next hundred. Some of these changes will be predictable, others not. Nonetheless, many diseases, have, at least in the west, have been beaten, or seemingly beaten, or, at least, subdued and kept in check. This is due to many factors which have militated against the severity and spread of disease; these include improvements to the way that life is lived – precautionary hygiene, nutrition, water quality, reduced overcrowding, improved living conditions – as well as more significant, interventionary measures, such as quarantining, antibiotic therapy, and, of course, vaccines.

Vaccine design and development is an inherently laborious process, but the programs and techniques outlined here have the potential to simplify the process greatly. The techniques described also have the potential to identify candidate proteins that would be overlooked by conventional experimentation. Reverse vaccinology has in particular proved effective in the discovery of antigenic subunit vaccines that would otherwise remain undiscovered.

It is sometimes difficult for outsiders to assess properly the relative merits of *in silico* vaccine design compared to mainstream experimental studies. The potential – albeit largely unrealised – is huge, but only if people are willing to take up the technology and use it appropriately. People's expectations of computational work are often largely unrealistic and highly tendentious. Some expect perfection, and are soon disappointed, rapidly becoming vehement critics. Others are highly critical from the start and are nearly impossible to reconcile with informatic methods. Neither appraisal is correct, however. Informatic methods do not replace, or even seek to replace, experimental work, only to help rationalise experiments, saving time and effort. They are slaves to the data used to generate them. They require a degree of intellectual effort equivalent in scale yet different in kind to that of so-called experimental science. The two disciplines – experimental and informatics – are thus complementary albeit distinct.

References

Andersen PH et al (2006) Prediction of residues in discontinuous B Cell epitopes using protein 3D structures. Protein Sci 15:2558–2567
Babiuk LA et al (2000) Nucleic acid vaccines: research tool or commercial reality. Vet Immunol Immunopathol 76:1–23
Babiuk LA et al (2003) Induction of immune responses by DNA vaccines in large animals. Vaccine 21:649–658
Bateman A et al (2000) The Pfam protein families database. Nucleic Acids Res 28:263–266
Bissantz C (2003) Conformational changes of G protein-coupled receptors during their activation by agonist binding. J Recept Signal Transduct Res 23:123–153
Blythe MJ, Flower DR (2004) Benchmarking B Cell epitope prediction: underperformance of existing methods. Protein Sci 14:246–248
Booth JS et al (2007) Innate immune responses induced by classes of CpG oligodeoxynucleotides in ovine lymph node and blood mononuclear cells. Vet Immunol Immunopathol 115(1–2): 24–34
Bulashevska A, Eils R (2006) Predicting protein subcellular locations using hierarchical ensemble of Bayesian classifiers based on Markov chains. BMC Bioinformatics. 7:298
Charoenvit Y, Goel N, Whelan M, Rosenthal KS, Zimmerman DH (2004a) CEL-1000 – a peptide with adjuvant activity for Th1 immune responses. Vaccine 22(19):2368–2373
Charoenvit Y et al (2004b) A small peptide (CEL-1000) derived from the beta-chain of the human major histocompatibility complex class II molecule induces complete protection against malaria in an antigen-independent manner. Antimicrob Agents Chemother 48:2455–2463
Christopoulos A, Kenakin TG (2002) protein-coupled receptor allosterism and complexing. Pharmacol Rev 54:323–374
Chvatchko Y, Hoogewerf AJ, Meyer A, Alouani S, Juillard P, Buser R, Conquet F, Proudfoot AE, Wells TN, Power CA (2000) A key role for CC chemokine receptor 4 in lipopolysaccharide-induced endotoxic shock. J Exp Med 191:1755–1764
Davies MN et al (2003) A novel predictive technique for the MHC class II peptide-binding interaction. Mol Med 9:220–225
Davies MN et al (2006) Statistical deconvolution of enthalpic energetic contributions to MHC-peptide binding affinity. BMC Struct Biol 6:5–17
De Groot AS (2006) Immunomics: discovering new targets for vaccines and therapeutics. Drug Discov Today 11:203–209
de Lisle GW, Wards BJ, Buddle BM, Collins DM (2005) The efficacy of live tuberculosis vaccines after presensitization with Mycobacterium avium. Tuberculosis (Edinb) 85(1–2):73–79
Delcher AL et al (1999) Improved microbial gene identification with GLIMMER. Nucleic Acids Res 27:4636–4641
Deng H, Liu R, Ellmeier W, Choe S, Unutmaz D, Burkhart M, Di Marzio P, Marmon S, Sutton RE, Hill CM, Davis CB, Peiper SC, Schall TJ, Littman DR, Landau NR (1996) Identification of a major co-receptor for primary isolates of HIV-1. Nature 381:661–666
Donnes P, Elofsson A (2002) Prediction of MHC class I binding peptides, using SVMHC. BMC Bioinformatics 3:25
Doytchinova IA, Flower DR (2007) VaxiJen: a server for prediction of protective antigens, tumour antigens and subunit vaccines. BMC Bioinformatics 8:4
Doytchinova IA et al (2005) Towards the chemometric dissection of peptide-HLA-A*0201 binding affinity: comparison of local and global QSAR models. J Comput Aided Mol Des 19:203–212
Ebo DG, Bridts CH, Stevens WJ (2008) IgE-mediated large local reaction from recombinant hepatitis B vaccine. Allergy 63(4):483–484
Emanuelsson O, Brunak S, von Heijne G, Nielsen H (2007) Locating proteins in the cell using TargetP, SignalP and related tools. Nat Protoc 2(4):953–971
Falquet L et al (2002) The PROSITE database. Nucleic Acids Res 30:235–238

Feng Y, Broder CC, Kennedy PE, Berger EA (1996) HIV-1 entry cofactor: functional cDNA cloning of a seven-transmembrane, G protein-coupled receptor. Science 272:872–877
Gether U, Asmar F, Meinild AK, Rasmussen SG (2002) Structural basis for activation of G-protein-coupled receptors. Pharmacol Toxicol 91:304–312
Girard MP, Osmanov SK, Kieny MP (2006) A review of vaccine research and development: the human immunodeficiency virus (HIV). Vaccine 24(19):4062–4081
Godessart N (2005) Chemokine receptors: attractive targets for drug discovery. Ann N Y Acad Sci 1051:647–657
Greenbaum JA, Andersen PH, Blythe M, Bui HH, Cachau RE, Crowe J, Davies MN et al (2007) Towards a consensus on datasets and evaluation metrics for developing B Cell epitope prediction tools. J Mol Recognit 20(2):75–82
Gupta RK (1998) Aluminum compounds as vaccine adjuvants. Adv Drug Deliv Rev 32:155–172
Guzmán E, Romeu A, Garcia-Vallve S (2008) Completely sequenced genomes of pathogenic bacteria: a review. Enferm Infecc Microbiol Clin 26(2):88–98
Harish N, Gupta R, Agarwal P, Scaria V, Pillai B (2006) DyNAVacS: an integrative tool for optimized DNA vaccine design. Nucleic Acids Res 34(Web Server issue):W264–W266
Hedrick JA, Zlotnik A (1996) Chemokines and lymphocyte biology. Curr Opin Immunol 8:343–347
Henry I, Sharp PM (2007) Predicting gene expression level from codon usage bias. Mol Biol Evol 24(1):10–12
Hung CF, Ma B, Monie A, Tsen SW, Wu TC (2008) Therapeutic human papillomavirus vaccines: current clinical trials and future directions. Expert Opin Biol Ther 8(4):421–439
Iellem A, Colantonio L, Bhakta S, Sozzani S, Mantovani A, Sinigaglia F, D'Ambrosio D (2001) Unique chemotactic response profile and specific expression of chemokine receptors CCR4 and CCR8 by CD4+CD25+ regulatory T cells. J Exp Med 194:847–854
Jardetzky TS et al (1996) Crystallographic analysis of endogenous peptides associated with HLADR1 suggests a common, polyproline II-like conformation for bound peptides. Proc Natl Acad Sci USA 93:734–738
Klinman DM et al (1997) Contribution of CpG motifs to the immunogenicity of DNA vaccines. J Immunol 158:3635–3639
Kulkarni-Kale U, Bhosle S, Kolaskar AS (2005) CEP: a conformational epitope prediction server. Nucleic Acids Res 33(Web Server issue):W168–W171
Lee AH et al (1997) Comparison of various expression plasmids for the induction of immune response by DNA immunization. Mol Cells 7:495–501
Lieberam I, Forster I (1999) The murine beta-chemokine TARC is expressed by subsets of dendritic cells and attracts primed $CD4^+$ T cells. Eur J Immunol 29:2684–2694
Liu W et al (2006) Quantitative prediction of mouse class I MHC peptide binding affinity using support vector machine regression (SVR) models. BMC Bioinformatics 7:182
Locati M, Murphy PM (1999) Chemokines and chemokine receptors: biology and clinical relevance in inflammation and AIDS. Annu Rev Med 50:425–440
Luster AD (1998) Chemokines – chemotactic cytokines that mediate inflammation. N Engl J Med 338:436–445
Maione D et al (2005) Identification of a universal Group B streptococcus vaccine by multiple genome screen. Science 309:148–150
McMurry J et al (2005) Analyzing *Mycobacterium tuberculosis* proteomes for candidate vaccine epitopes. Tuberculosis (Edinb) 85:95–105
Merino EF, Fernandez-Becerra C, Durham AM, Ferreira JE, Tumilasci VF, d'Arc-Neves J, da Silva-Nunes M, Ferreira MU, Wickramarachchi T, Udagama-Randeniya P, Handunnetti SM, Del Portillo HA (2006) Multi-character population study of the vir subtelomeric multigene superfamily of Plasmodium vivax, a major human malaria parasite. Mol Biochem Parasitol 149(1):10–16
Mok BW, Ribacke U, Winter G, Yip BH, Tan CS, Fernandez V, Chen Q, Nilsson P, Wahlgren M (2007) Comparative transcriptomal analysis of isogenic Plasmodium falciparum clones of distinct antigenic and adhesive phenotypes. Mol Biochem Parasitol 151(2):184–192
Montigiani S et al (2002) Genomic approach for analysis of surface proteins in Chlamydia pneumoniae. Infect Immun 70:368–379

Noguchi H et al (2002) Hidden Markov model-based prediction of antigenic peptides that interact with MHC class II molecules. J Biosci Bioeng 94:264–270
O'Hagan DT, MacKichan ML, Singh M (2001) Recent developments in adjuvants for vaccines against infectious diseases. Biomol Eng 18(3):69–85
Ou HY et al (2004) GS-Finder: a program to find bacterial gene start sites with a self-training method. Int J Biochem Cell Biol 36:535–544
Palma-Carlos AG, Santos AS, Branco-Ferreira M, Pregal AL, Palma-Carlos ML, Bruno ME, Falagiani P, Riva G (2006) Clinical efficacy and safety of preseasonal sublingual immunotherapy with grass pollen carbamylated allergoid in rhinitic patients. A double-blind, placebo-controlled study. Allergol Immunopathol (Madr) 34(5):194–198
Pizza M et al (2000a) Identification of vaccine candidates against serogroup B meningococcus by whole-genome sequencing. Science 287:1816–1820
Pizza M et al (2000b) Whole genome sequencing to identify vaccine candidates against serogroup B meningococcus. Science 287:1816–1820
Ponomarenko JV, Bourne PE (2007) Antibody-protein interactions: benchmark datasets and prediction tools evaluation. BMC Struct Biol 7:64
Rammensee H et al (1999) SYFPEITHI: database for MHC ligands and peptide motifs. Immunogenetics 50:213–219
Rey S, Acab M, Gardy JL, Laird MR, deFays K, Lambert C, Brinkman FS (2005) PSORTdb: a protein subcellular localization database for bacteria. Nucleic Acids Res 33(Database issue):D164–D168
Rombel IT et al (2003) ORF-FINDER: a vector for high-throughput gene identification. Gene 282:33–41
Ross BC et al (2001) Identification of vaccine candidate antigens from a genomic analysis of Porphyromonas gingivalis. Vaccine 19:4135–4142
Salomon J, Flower DR (2006) Predicting Class II MHC-Peptide binding: a kernel based approach using similarity scores. BMC Bioinformatics 7:501
Schellhammer I, Rarey M (2004) FlexX-Scan: fast, structure-based virtual screening. Proteins 57(3):504–517
Schijns VE (2003) Mechanisms of vaccine adjuvant activity: initiation and regulation of immune responses by vaccine adjuvants. Vaccine 21:829–831
Schreiber A, Humbert M, Benz A, Dietrich U (2005) 3D-Epitope-Explorer (3DEX): localization of conformational epitopes within three-dimensional structures of proteins. J Comput Chem 26(9):879–887
Servant F et al (2002) ProDom: automated clustering of homologous domains. Brief Bioinform 3:246–251
Singh M, O'Hagan DT (2002) Recent advances in vaccine adjuvants. Pharm Res 19:715–728
Singh M, Srivastava I (2003) Advances in vaccine adjuvants for infectious diseases. Curr HIV Res 1:309–320
Stills HF Jr (2005) Adjuvants and antibody production: dispelling the myths associated with Freund's complete and other adjuvants. ILAR J 46:280–293
Tettelin H et al (2000) Complete genome sequence of Neisseria meningitidis serogroup B strain MC58. Science 287:1809–1815
Tongchusak S, Brusic V, Chaiyaroj SC (2008) Promiscuous T cell epitope prediction of Candida albicans secretory aspartyl proteinase family of proteins. Infect Genet Evol 8(4):467–473
Uchijima M et al (1998) Optimization of codon usage of plasmid DNA vaccine is required for the effective MHC class I-restricted T Cell responses against an intracellular bacterium. J Immunol 161:5594–5599
Vekemans J, Ballou WR (2008) Plasmodium falciparum malaria vaccines in development. Expert Rev Vaccines 7(2):223–240
Vivona S, Bernante F, Filippini F.(2006) NERVE: new enhanced reverse vaccinology environment. BMC Biotechnol 6:35
Wan S et al (2004) Large-scale molecular dynamics simulations of HLA-A*0201 complexed with a tumor-specific antigenic peptide: can the alpha3 and beta2m domains be neglected? J Comput Chem 25:1803–1813

Wan J et al (2006) SVRMHC prediction server for MHC-binding peptides. BMC Bioinformatics 7:463

Winnenburg R, Urban M, Beacham A, Baldwin TK, Holland S, Lindeberg M, Hansen H, Rawlings C, Hammond-Kosack KE, Köhler J (2008) PHI-base update: additions to the pathogen host interaction database. Nucleic Acids Res 36(Database issue):D572–D576

Wizemann TM et al (2001) Use of a whole genome approach to identify vaccine molecules affording protection against *Streptococcus pneumoniae* infection. Infect Immun 69:1593–1598

Xu ZL et al (2001) Optimization of transcriptional regulatory elements for constructing plasmid vectors. Gene 272:149–156

Yang J, Chen L, Sun L, Yu J, Jin Q (2008) VFDB 2008 release: an enhanced web-based resource for comparative pathogenomics. Nucleic Acids Res 36(Database issue):D539–D542

The Immuno Polymorphism Database

James Robinson and Steven G.E. Marsh

Introduction

The Immuno Polymorphism Database is a set of specialist databases related to the study of polymorphic genes in the immune system. IPD currently consists of four databases: IPD-KIR, which contains the allelic sequences of Killer-cell Immunoglobulin-like Receptors (KIR); IPD-MHC, which is a database of sequences of the major histocompatibility complex (MHC) of different species; IPD-HPA, which has alloantigens expressed only on platelets; and IPD-ESTDAB, which provides access to the European Searchable Tumour Cell-Line Database (ESTDAB), a cell bank of immunologically characterised melanoma cell lines.

The study of the immune system constitutes many different complex areas of research. The aim of the IPD Database project is to provide a centralised resource for information pertaining to polymorphic genes of the immune system by coupling the expertise of various different research groups or nomenclature committees with the informatics experience of the HLA Informatics Group at the Anthony Nolan Research Institute. The individual experts or nomenclature committees are established within their own fields, each has a role in assessing the quality and validity of new data submitted to their own section of the database. This may be in the identification and naming of new alleles based on the submission of new sequences to generalist databases like the European Molecular Biology Laboratory's nucleotide database (EMBL), the National Center for Biotechnology Information's GenBank and the DNA DataBank of Japan (DDBJ) (Benson et al. 2008; Cochrane et al. 2008; Sugawara et al. 2008; Tateno 2008). Or in the collation and validation of data from a variety of different cell characterisation methods, such as the cases for the IPD-ESTDAB database. One advantage of using a centralised system is the ability to share or reuse elements of database structure, when dealing with similar datasets.

S.G.E. Marsh (✉)
The Anthony Nolan Research Institute, Royal Free Hospital, Pond Street, Hampstead, NW3 2QG, London, UK
UCL Cancer Institute, Royal Free Campus, Pond Street, Hampstead, NW3 2QG, London, UK
e-mail: marsh@ebi.ac.uk

D.R. Flower et al. (eds.), *Bioinformatics for Immunomics*, Immunomics Reviews 3,
DOI 10.1007/978-1-4419-0540-6_2, © Springer Science+Business Media, LLC 2010

Much of the database structure of IPD-KIR and –MHC sections is shared with the IMGT/HLA database (Robinson and Marsh 2006). This has also enabled cross-database implementation of some of the core tools, particularly those for data analysis, submission and retrieval.

IPD Projects

IPD-MHC

The Major Histocompatibility Complex (MHC) sequences of many species have been reported in the literature and are represented in the generalist sequence databases. For some species or related species groups, such as the bovines, non-human primates (NHP) and dogs, there have been efforts to use a standardised nomenclature system and establish comprehensive datasets (Kennedy et al. 1999; Klein et al. 1990; Stear et al. 1990). The availability of these datasets for use by other groups has often been limited by a lack of informatics resources available to the researchers compiling the datasets. The aim of the IPD-MHC database is to address this issue and provide a centralised database which will facilitate the comparative analysis of these sequences which are highly conserved between different closely related species (Parham 1999). In addition the formation of the International Society of Animal Genetics (ISAG)/ International Union of Immunological Society (IUIS)-Veterinary Immunology Committee (VIC) Comparative MHC Nomenclature Committee, which brings together representatives of many nomenclature committees covering different species, will aid in the establishment of standardised nomenclature practices across species (Ellis et al. 2006) (Fig. 1).

For each species, there are differences in the spectrum of data covered, but all sections provide the core nomenclature pages and sequence alignments. The nomenclature and alignments follow a similar structure to that of the IPD-KIR section, and the same basic tools are used in both sections. Some nomenclature committees may provide additional information but the core components of any nomenclature reported are the allele names, accession numbers and publications.

Currently the IPD-MHC sequence alignments are limited to species-specific alignments; however, we are working to allow cross-species alignments and the inclusion of human sequences from the IMGT/HLA Database (Ruiz et al. 2000) for comparative purposes. The alignments for non-human primates can use a human HLA sequence as a reference sequence, but this is only a single sequence in each alignment. The IPD-MHC Database also contains a submission tool for online submission of new and confirmatory sequences to the appropriate nomenclature committee.

The first release of the IPD-MHC database involved the work of groups specialising in non-human primates, canines and felines, and incorporated all data previously available in the IMGT/MHC database (Robinson et al. 2003). A subsequent release added cattle, fish, rat, sheep and swine. In addition recent developments within the IPD-MHC NHP section now allow online queries of the database through an ‘Allele Query Tool’. This is designed to allow the user to retrieve the

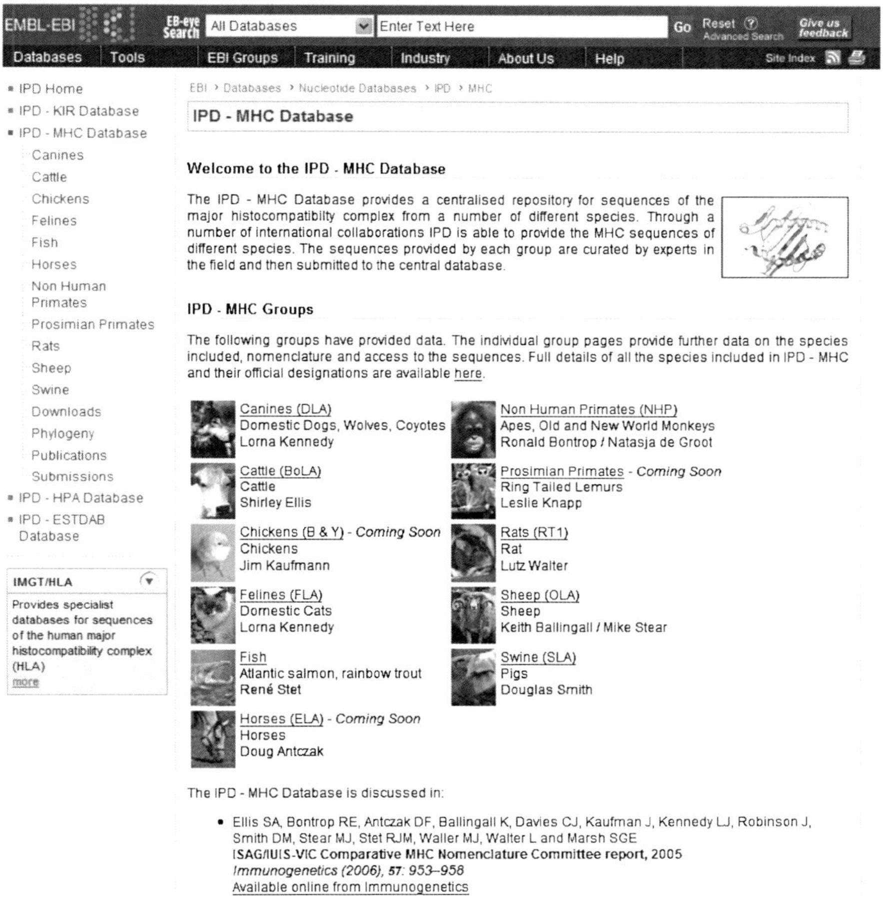

Fig. 1 The IPD-MHC Home Page**.** The IPD-MHC Home Page provides access to data on the major histocompatibility complex of different species as well as links to the other IPD component databases: IPD-KIR, which contains the allelic sequences of Killer-cell Immunoglobulin-like Receptors (KIR); IPD-HPA, which has alloantigens expressed only on platelets; and IPD-ESTAB, which provides access to the European Searchable Tumour Cell-Line Database (ESTDAB)

full sequence and information pertaining to any officially named NHP allele. The tool provides a simple to use interface for retrieving allele information. The 'Allele Query Tool' requires only a search term in order to retrieve a report on any allele in the database. The search tool allows different resolutions of the allele name allow flexibility in searches. The chimpanzee allele, *Patr*-A*0101, can be searched for at the species level by entering '*Patr*', at the gene level by entering '*Patr*-A' or at the allele level by entering '*Patr*-A*0101' in the search tool. This facility can be exploited to gain a list of all alleles at any locus or species by entering either '*Species-locus*' or '*Species*' in the box to retrieve the full list. The output of the search lists all the allele names that correspond to the search term provided; these then provide a hypertext link to the full entry for each allele. The output provided for each allele includes the official allele designation, previously used designations and the unique IPD-MHC accession number. Other information provided includes the date that the allele was named and its current status (as some allele designations have been deleted). Links to all component EMBL/GenBank/DDBJ entries are also included. Any published references are also included with, wherever possible, a link to the PubMed entry for that citation. The PubMed link provides an online version of the abstract as well as links to other citations by the author and to similar papers. The final section of the output details the official nucleotide and protein sequence as well as where available any genomic sequence for that allele. Future releases are planned to include chicken, horse, prosimian and piniped (seal and sea lion) sequences.

IPD-KIR

The Killer-cell Immunoglobulin-like Receptors (KIR) are members of the immunoglobulin super family (IgSF) formerly called Killer-cell Inhibitory Receptors. KIRs have been shown to be highly polymorphic both at the allelic and haplotypic levels (Garcia et al. 2003). They are composed of two or three Ig-domains, a transmembrane region and cytoplasmic tail, which can in turn be short (activatory) or long (inhibitory). The Leukocyte Receptor Complex (LRC), which encodes KIR genes, has been shown to be polymorphic, polygenic and complex in a manner similar to the MHC, because of the complexity in the KIR region and KIR sequences a KIR Nomenclature Committee was established in 2002, to undertake the naming of KIR allele sequences. The first KIR Nomenclature report was published in 2002 (Marsh et al. 2003), which coincided with the first release of the IPD-KIR database. To aid in the analysis and naming of new alleles, an online submission tool is provided on the IPD-KIR website. New alleles are added to the database once they have successfully completed the submission procedure and their quality assured. Periodic new releases of the database contain newly submitted sequences together with conformations and extensions of those already available.

The KIR Nomenclature Committee is also involved in the naming of the complex haplotypes and genotypes currently seen in KIR research. Proposals for such a nomenclature have been published, but as yet this nomenclature has not been implemented, although it is planned to include this data once available. The online tools available for IPD-KIR include allele queries, sequence alignments and cell queries. As the database is based on the work of a nomenclature committee, the website includes links to a portable document format (PDF) file of recent nomenclature reports. From the data contained within these reports the database is also able to provide individual allele reports (Fig. 2). These pages contain the official allele name, any previous designations, the EMBL, GenBank, or DDBJ accession number(s) and a reference linked wherever possible to the PubMed abstract.

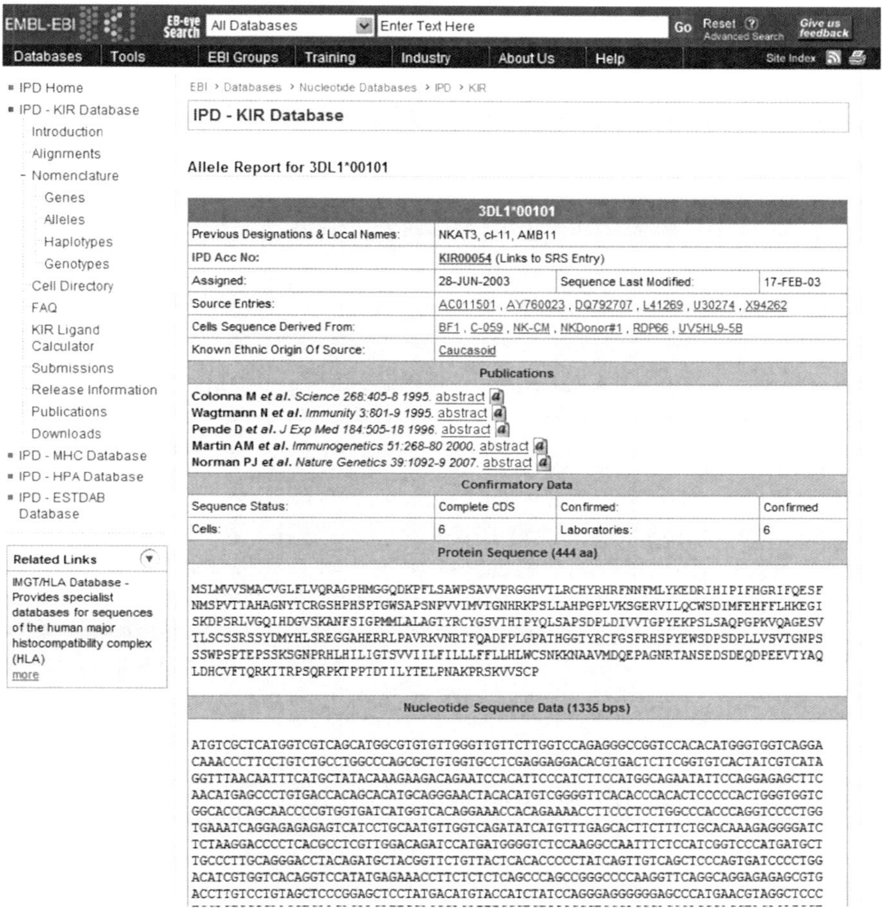

Fig. 2 IPD-KIR Allele Entry. From the data contained within the nomenclature reports the database is able to provide individual allele reports. The report shown shows part of the KIR3DL1*00101 entry. The underlined text is a link to further information on the web both within IPD and in external sources like PubMed

Where possible additional details on the source of sequence are also provided. This source material is normally in the form of a cell line or DNA from which each allele in the database was isolated and characterised. The information contained within this dataset can be searched independently from the allele data. The cell query tool is used to interrogate this accompanying cell database. The interface can be used without prior knowledge of which alleles are linked to the cells or from the allele reports. The interface allows the user to search on known cell fields. The cells all have a primary name and accession number that are unique within the database. As some cells are sequenced by different groups, or certain names are repeated, the database also contains a list of aliases for each cell. These aliases are automatically searched whenever the cell name field is queried. Other cell fields that are available for searching include HLA and KIR typing, serology, ethnic origin, geographical location.

Recent additions to the tools available from the IPD-KIR site include a KIR Ligand Calculator (Yun et al. 2007). The ligand calculator allows the user to define which KIR ligands are present in a transplant setting based on the HLA typing of the donor and patient. This is because some recent haematopoietic stem cell transplant strategies that have been based on KIR-ligand mismatch to predict NK cell alloreactivity have resulted in less relapse, less GvHD and better overall survival in patients with Acute Myeloid Leukaemia (AML) (Gumperz et al. 1997; Khakoo et al. 2004; Ruggeri et al. 1999). The KIR-ligands are HLA molecules that can be grouped into three major categories based on the amino acid sequence determining the KIR-binding epitope in HLA-C and HLA-B molecules.

The typing of KIRs is dependant on up-to-date lists of alleles and primers, and many typing laboratories have spreadsheets detailing probe hit patterns for different alleles. Each time a new database was released it was necessary to update these ever-expanding lists. The Probe & Primer Search Tool allows users to enter a list of primer sequences and the tool will search the known coding sequences for these and report any matches in file format suitable for cutting and pasting into existing spreadsheets. The tool is currently limited to coding sequence, but as the number of genomic sequences in the database expands the tool can be modified to search these regions as well.

The IPD-KIR database has also been expanded to include the KIR sequences from other species, most recently work has begun on including the sequences of KIR alleles found in Rhesus Macaques (*Macaca mulatta*) (Bimber and O'Connor 2008).

Sequence Alignments in IPD-MHC and IPD-KIR

Within each IPD section allele sequences may differ from each other by as little as a single nucleotide. As discussed in Chapter 3. The IMGT/HLA Database (Robinson and Marsh 2008), these alignments allow a visual interpretation of sequence similarity, so that polymorphic positions can easily be identified and motifs found in multiple alleles are easily identified. The sequence alignments are available via a link from the section homepage. The sequence alignment tool uses the same basic interface for both IPD-KIR database and IPD-MHC, see Fig. 3. The interface provided lets the user define a number of key variables for the alignments, before producing an online output, which can be printed or downloaded. The first step in any alignment is to select the locus of interest. The tool provides a drop-down list of all loci. Where more than one species is available within a particular section, the interface will contain an additional option to define the species of interest. The selection of a locus automatically updates the list of features, which can be aligned, as well as the default reference sequence used for the alignment. The types of feature available for alignment are the nucleotide coding sequence and individual exons, the signal peptide, mature protein and full-length protein sequence. The alignment tool options also allow the user to display a subset of alleles of a particular locus, omit alleles unsequenced for a particular region and align against a particular reference

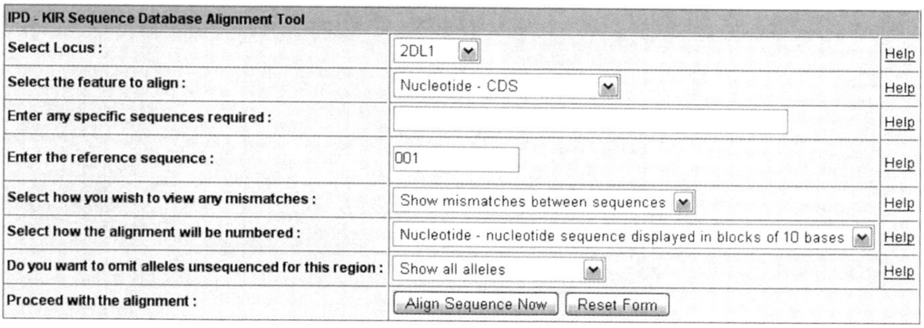

Fig. 3 Alignment interface. The alignment interface provides a user-friendly method of viewing sequence alignments with output options easily selected

or consensus sequence. The alignment tool uses standard formatting conventions for the display of sequence alignments (Fig. 4). The alignment tool does not perform a sequence alignment each time it is used, but it extracts pre-aligned sequences, allowing for faster access.

The alignments adhere to a number of conventions for displaying evolutionary events and numbering. The numbering of the alignments is based upon the sequence of the reference allele. For a nucleotide sequence, the A of the initiation Methionine codon is denoted nucleotide +1 and the nucleotide 5′ to +1 is numbered −1. There is no nucleotide zero (0). All numbering is based on the ATG of the reference sequence. If a nucleotide sequence is displayed in codons, then the protein numbering is applied.

For amino acid based alignments, the first codon of the mature protein, after cleavage of the signal sequence, is labelled codon 1 and the codon 5′ to this is numbered −1. In all sequences the following conventions are used. Where identity to the reference sequence is present the base will be displayed as a hyphen (-). Non-identity to the reference sequence is shown by displaying the appropriate base at that position. Where an insertion or deletion has occurred this is represented by a period (.). If the sequence is unknown at any point in the alignment, this is to be represented by an asterisk (*). In protein alignments for null alleles, the 'Stop' codons are represented by an X and further sequence following the termination codon is not displayed and appears blank. The flexibility of the alignment tool means that unlike previous alignments you can now display a small subset of sequences against an allele of your choice, using a number of display options. If a user would rather use other existing software to produce their own alignments then the FTP directory contains files in popular formats for them to download and import.

IPD-HPA

Human Platelet Antigens (HPA) are alloantigens expressed on platelets, specifically on platelet membrane glycoproteins. These platelet-specific antigens

a

```
                                    1           5                        10                      15                    20
Ovar-DRB1*0101   CTG GCC TGG GCC AGG AAG ATC CAA CCA CAT TTC TTG GAG TAT ACT AAG AAA GAG TGT CGT TTC TCC AAC GGG ACG
Ovar-DRB1*0201   --- --- --- --- --- --- --- --- --- --- --- --- --- --- T-- -C- -GC --- --- -A- --- -T- --- --- ---
Ovar-DRB1*0301   --- --- --- --- --- --- --- --- --- --- --- --- --- --- --- --- --- --- --- --- --- --- --- --- ---
Ovar-DRB1*0302   *** *** *** *** *** *** *** *** **- --- --- --- --- --- --- --- --- --- --- --- --- --- --- --- ---
Ovar-DRB1*0303   --- --- --- --- --- --- --- --- --- --- --- --- --- --- --- --- --- --- --- --- --- --- --- --- ---
Ovar-DRB1*0401   *** *** *** *** *** *** *** *** **- --- --- --- --- --- CA- --- -GC --- --- --- --- --- --- --- ---
Ovar-DRB1*0501   --- --- --- --- --- --- --- --- --- --- --- --- --- --- G-- --- -GC --- --- --- --- -T- --- --- ---
Ovar-DRB1*0601   --- --- --- --- --- --- --- --- --- --- --- --- --- --- G-- --- -GC --- --- --- --- -T- --- --- ---
```

b

```
                        110        120        130        140        150        160        170        180        190        200
Ovar-DRB1*0101   CACATTTCTT GGAGTATACT AAGAAAGAGT GTCGTTTCTC CAACGGGACG GAGCGGGTGC GGTTCCTGGA CAGATACTTC CATAATGGAG AAGAGACCCT
Ovar-DRB1*0201   ---------- -------T-- -C--GC---- ---A-----T ---------- ---------- ---------- ---------- T--------- ----------
Ovar-DRB1*0301   ---------- ---------- ---------- ---------- ---------- ---------- ---------- ---------- ---------- -----TA-GC
Ovar-DRB1*0302   *--------- ---------- ---------- ---------- ---------- ---------- ---------- ---------- T--------- -----TA-G-
Ovar-DRB1*0303   ---------- ---------- ---------- ---------- ---------- ---------- ---------- ---------- T--------- -----TA-GC
Ovar-DRB1*0401   *--------- -------CA- ----GC---- ---------- ---------- ---------- ---A------ ---------- T--------- -----TA-G-
Ovar-DRB1*0501   ---------- -------G-- ----GC---- ---------T ---------- ---------- ---------- A--------- T--------- ----------
Ovar-DRB1*0601   ---------- -------G-- ----GC---- ---------T ---------- ---------- ---------- A--------- T--------- -----TA-G-
```

c

```
                        10         20         30         40         50         60         70         80         90        100
Patr-A*0101      MAVMPPRTLL LLLSGALALT QTWAGSHSMR YFFTSVSRPG RGEPRFIAVG YVDDTQFVRF DSDAASQRME PRAPWIEQEG PEYWDQETRS AKAHSQTDRV
Patr-A*0201      ---------- ---------- ---------- ---------- ---------- ---------- ---------- ---------- -----E---- ----------
Patr-A*0301      ---------- ---------- ---------- --Y------- ---------- ---------- ---------- ---------- ---------N M--SA-----
Patr-A*0302      ---------- ---------- ---------- --Y------- ---------- ---------- ---------- ---------- ---------N M--SA-----
Patr-A*0401      ----A----V ---------- ---------- --S------- ---------- ---------- ---------- ---------- -----E---- V--SA-----
Patr-A*0402      ----A----V ---------- ---------- --S------- ---------- ---------- ---------- ---------- -----E---- V--SA-----
Patr-A*0404      **--A----V ---------- ---------- --S------- ---------- ---------- ---------- ---------- -----E---- V--QA-----
Patr-A*0501      ----A----V ---------- ---------- --S------- ---------- ---------- ---------- ---------- -----E---- V--FA-----
```

d

```
                        10         20         30         40         50         60         70         80         90        100
 HLA-A*01010101  MAVMAPRTLL LLLSGALALT QTWAGSHSMR YFFTSVSRPG RGEPRFIAVG YVDDTQFVRF DSDAASQKME PRAPWIEQEG PEYWDQETRN MKAHSQTDRA
Patr-A*0101      ----P----- ---------- ---------- ---------- ---------- ---------- -------R-- ---------- ---------S A--------V
Patr-A*0201      ----P----- ---------- ---------- ---------- ---------- ---------- -------R-- ---------- -----E---S A--------V
Patr-A*0301      ----P----- ---------- ---------- --Y------- ---------- ---------- -------R-- ---------- ---------- ---SA----V
Patr-A*0302      ----P----- ---------- ---------- --Y------- ---------- ---------- -------R-- ---------- ---------- ---SA----V
Patr-A*0401      ---------V ---------- ---------- --S------- ---------- ---------- -------R-- ---------- -----E---S V--SA----V
Patr-A*0402      ---------V ---------- ---------- --S------- ---------- ---------- -------R-- ---------- -----E---S V--SA----V
Patr-A*0404      **-------V ---------- ---------- --S------- ---------- ---------- -------R-- ---------- -----E---S V--QA----V
Patr-A*0501      ---------V ---------- ---------- --S------- ---------- ---------- -------R-- ---------- -----E---S V--FA----V
```

Fig. 4 Alignment formats available from IPD-MHC. The examples shown are all alignments of DRB alleles from different species. In these alignments a *dash* indicates identity to the reference sequence and an *asterisk* denotes an unsequenced base. The first two alignments (A and B) show the nucleotide sequences of Sheep (*Ovis aries*) DRB1 alleles using different display parameters. The second set shows some Chimpanzees (*Pan troglodytes*) Patr-A protein sequences. The same sets of alleles are used for both C and D, but in D the Chimpanzee sequences are aligned against a human reference sequence (HLA-A*01010101)

are immunogenic and can result in pathological reactions to transfusion therapy. The HPA nomenclature system was adopted in 1990 (von dem Borne and Decary 1990a, b) to overcome problems with the previous nomenclature. Since then, more antigens have been described and the molecular basis of many has been resolved. As a result the nomenclature was revised in 2003 (Metcalfe et al. 2003) and included in the IPD project. The IPD-HPA section contains nomenclature information and additional background material. The different genes in the HPA system have not been sequenced to the same level as some of the other projects, and so currently only single nucleotide polymorphisms (SNP) are used to determine alleles. This information is presented in a grid of SNP for each gene. The HPA Nomenclature Committee hopes to expand this to provide full sequence alignments when possible. The IPD-HPA section also provides data on the frequency of different HPA alleles in a number of populations. These tables contain allele frequencies as well as the ethnic origins of the samples, typing methodology and relevant publications. This table is regularly updated and is now considered one of the main resources for HPA frequency data (Fig. 5).

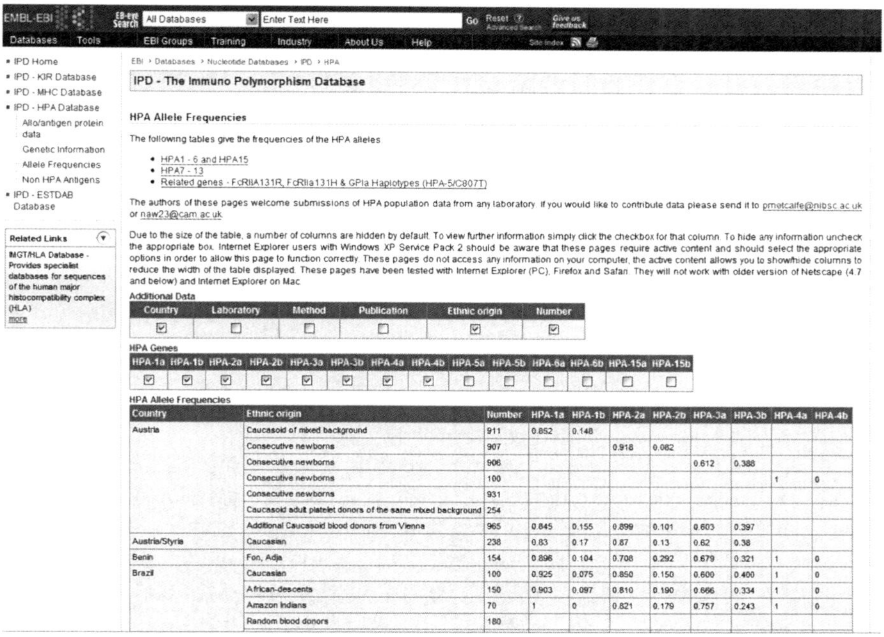

Fig. 5 IPD-HPA frequency data. The IPD-HPA database contains frequency data on different HPA alleles. A selection of data is shown in this figure. Further columns can be added to the table by selecting the appropriate column from the menus above the main report. The table already selected are identified by a tick mark in the relevant box

IPD-ESTDAB

IPD-ESTDAB is a database of immunologically well-characterised melanoma cell lines. The database works in conjunction with the European Searchable Tumour Cell Line Database (ESTDAB) cell bank, which is housed in Tübingen, Germany, and provides access to the immunologically characterised tumour cells (Pawelec and Marsh 2006). The ESTDAB consortium is made up of seven laboratories from countries around the European Union, with each lab responsible for the generation of data on a prearranged set of immunological or genetic markers that reflect that laboratory's expertise and technical specialities. The central facility and physical location of the cell bank was established at the Centre for Medical Research (ZMF) of the University of Tübingen in Germany, where cells lines were gathered from a variety of sources around Europe, Australia and the United States of America. Since the project began, cells have been acquired as they become available; this has meant in particular that a number of cell lines have been sourced from melanoma samples collected from patients entered into clinical immunotherapy trials associated with ESTDAB's sister project, OISTER (Outcome and Impact of Specific Treatment in European Research on Melanoma, http://www.dkfz.de/oister/). Consequently ESTDAB now provides access to many more cell lines of melanoma origin than are available from other non-specialist cell banks. The IPD-ESTDAB section of the website provides an online search facility for cells stored in this cell bank. This enables investigators to identify cells possessing specific parameters important for studies of immunity, immunogenetics, gene expression, metastasis, response to chemotherapy and other tumour biological experimentation. The search tool allows for searches based on a single parameter, or clusters of parameters on over 250 different markers for each cell. The detailed reports produced can then be used to identify cells of interest, which can then be obtained from the cell bank. Some elements of the design of the ESTDAB database are borrowed from that used in the IMGT/HLA Database.

Discussion

The IPD project provides a new resource for those interested in the study of polymorphic sequences of the immune system. By accommodating related systems in a single database, data can be made available in common formats aiding use and interpretation. As the projects grow and more sections are added, the benefit of having expertly curated sequences from related areas stored in a single location will become more apparent. This is particularly true of the IPD-MHC project, which already contains 2,270 sequences from over 50 different species, where cross-species studies will be able to utilise the high-quality sequences provided by the different nomenclature committees in a common format, ready for use. The initial release of the IPD Database contained only four sections and a small number of tools; however, as the database grows and more sections and species are added, more tools will be added to the website. We plan to use the existing database structures to house data for new sections of the IPD project as they become available. Data will also be made available in different formats to download from the website, FTP server and included into SRS, BLAST and FASTA search engines at the European Bioinformatics Institute (Harte et al. 2004).

Acknowledgements We would like to thank Peter Parham, Libby Guethlein, Christian Garcia and Chrissy Roberts for their work on IPD-KIR; our IPD-MHC collaborators Shirley Ellis, Lorna Kennedy, Ronald Bontrop, Natasja de Groot, Lutz Walter, Douglas Smith, Keith Ballingall, Mike Stear and René Stet; Nick Watkins and Paul Metcalfe for IPD-HPA and Graham Pawelec for ESTDAB. Our thanks also go to Matthew Waller and Sylvie Fail for their work on the tools and processing of submissions for the various databases and finally Peter Stoehr at the EBI for assisting in the continued collaboration between our group and the EBI.

Appendix: Access and Contact

IPD Homepage	http://www.ebi.ac.uk/ipd/
IPD-KIR Homepage	http://www.ebi.ac.uk/ipd/kir/
IPD-MHC Homepage	http://www.ebi.ac.uk/ipd/mhc/
IPD-HPA Homepage	http://www.ebi.ac.uk/ipd/hpa/
IPD-ESTDAB Homepage	http://www.ebi.ac.uk/ipd/estdab/

Contact: ipd@ebi.ac.uk

If you are interested in contributing to the project, there are specific guidelines for the inclusion of new sections, and interested parties should contact Prof. SGE Marsh, marsh@ebi.ac.uk for further information.

References

Benson DA, Karsch-Mizrachi I, Lipman DJ, Ostell J, Wheeler DL (2008) GenBank. Nucleic Acids Res 36:D25–D30

Bimber B, O'Connor DH (2008) KIRigami: the case for studying NK cell receptors in SIV+ macaques. Immunol Res 40:235–243

Cochrane G, Akhtar R, Aldebert P, Althorpe N, Baldwin A, Bates K, Bhattacharyya S, Bonfield J, Bower L, Browne P, Castro M, Cox T, Demiralp F, Eberhardt R, Faruque N, Hoad G, Jang M, Kulikova T, Labarga A, Leinonen R, Leonard S, Lin Q, Lopez R, Lorenc D, McWilliam H, Mukherjee G, Nardone F, Plaister S, Robinson S, Sobhany S, Vaughan R, Wu D, Zhu W, Apweiler R, Hubbard T, Birney E (2008) Priorities for nucleotide trace sequence and annotation data capture at the Ensembl Trace Archive and the EMBL Nucleotide Sequence Database. Nucleic Acids Res 36:D5–D12

Ellis SA, Bontrop RE, Antczak DF, Ballingall K, Davies CJ, Kaufman J, Kennedy LJ, Robinson J, Smith DM, Stear MJ, Stet RJ, Waller MJ, Walter L, Marsh SGE (2006) ISAG/IUIS-VIC Comparative MHC Nomenclature Committee report, 2005. Immunogenetics 57:953–958

Garcia CA, Robinson J, Guethlein LA, Parham P, Madrigal JA, Marsh SGE (2003) Human KIR sequences 2003. Immunogenetics 55:227–239

Gumperz JE, Barber LD, Valiante NM, Percival L, Phillips JH, Lanier LL, Parham P (1997) Conserved and variable residues within the Bw4 motif of HLA-B make separable contributions to recognition by the NKB1 killer cell-inhibitory receptor. J Immunol 158:5237–5241
Harte N, Silventoinen V, Quevillon E, Robinson S, Kallio K, Fustero X, Patel P, Jokinen P, Lopez R (2004) Public web-based services from the European Bioinformatics Institute. Nucleic Acids Res 32:W3–W9
Kennedy LJ, Altet L, Angles JM, Barnes A, Carter SD, Francino O, Gerlach JA, Happ GM, Ollier WE, Polvi A, Thomson W, Wagner JL (1999) Nomenclature for factors of the dog major histocompatibility system (DLA), 1998. First report of the ISAG DLA Nomenclature Committee. International Society for Animals Genetics. Tissue Antigens 54:312–321
Khakoo SI, Thio CL, Martin MP, Brooks CR, Gao X, Astemborski J, Cheng J, Goedert JJ, Vlahov D, Hilgartner M, Cox S, Little AM, Alexander GJ, Cramp ME, O'Brien SJ, Rosenberg WM, Thomas DL, Carrington M (2004) HLA and NK cell inhibitory receptor genes in resolving hepatitis C virus infection. Science 305:872–874
Klein J, Bontrop RE, Dawkins RL, Erlich HA, Gyllensten UB, Heise ER, Jones PP, Parham P, Wakeland EK, Watkins DI (1990) Nomenclature for the major histocompatibility complexes of different species: a proposal. Immunogenetics 31:217–219
Marsh SGE, Parham P, Dupont B, Geraghty DE, Trowsdale J, Middleton D, Vilches C, Carrington M, Witt C, Guethlein LA, Shilling H, Garcia CA, Hsu KC, Wain H (2003) Killer-cell immunoglobulin-like receptor (KIR) nomenclature report, 2002. Immunogenetics 55:220–226
Metcalfe P, Watkins NA, Ouwehand WH, Kaplan C, Newman P, Kekomaki R, De Haas M, Aster R, Shibata Y, Smith J, Kiefel V, Santoso S (2003) Nomenclature of human platelet antigens. Vox Sang 85:240–245
Parham P (1999) Virtual reality in the MHC. Immunol Rev 167:5–15
Pawelec G, Marsh SG (2006) ESTDAB: a collection of immunologically characterised melanoma cell lines and searchable databank. Cancer Immunol Immunother 55(6):623–627
Robinson J, Marsh SGE (2006) Immunoinformatics: predicting immunogenicity in silico. In: Flower D (ed) The IMGT/HLA Database. Humana, Totowa, pp 43–60
Robinson J, Marsh SGE (2008) In: Davies MN, Ranganathan S, Flower DR (eds) Bioinformatics for immunomics, Springer, pp 33–45
Robinson J, Waller MJ, Parham P, de Groot N, Bontrop R, Kennedy LJ, Stoehr P, Marsh SGE (2003) IMGT/HLA and IMGT/MHC: sequence databases for the study of the major histocompatibility complex. Nucleic Acids Res 31:311–314
Ruggeri L, Capanni M, Casucci M, Volpi I, Tosti A, Perruccio K, Urbani E, Negrin RS, Martelli MF, Velardi A (1999) Role of natural killer cell alloreactivity in HLA-mismatched hematopoietic stem cell transplantation. Blood 94:333–339
Ruiz M, Giudicelli V, Ginestoux C, Stoehr P, Robinson J, Bodmer J, Marsh SGE, Bontrop R, Lemaitre M, Lefranc G, Chaume D, Lefranc MP (2000) IMGT the international ImMunoGeneTics database. Nucleic Acids Res 28:219–221
Stear MJ, Pokorny TS, Fryda-Bradley S, Lie O, Bull RW (1990) Genetic analysis of the antigens defined at the third international BoLA workshop. J Immunogenet 17:21–28
Sugawara H, Ogasawara O, Okubo K, Gojobori T, Tateno Y (2008) DDBJ with new system and face. Nucleic Acids Res 36:D22–D24
Tateno Y (2008) [International collaboration among DDBJ EMBL Bank and GenBank]. Tanpakushitsu Kakusan Koso 53:182–189
von dem Borne AE, Decary F (1990a) ICSH/ISBT Working Party on platelet serology. Nomenclature of platelet-specific antigens. Vox Sang 58:176
von dem Borne AE, Decary F (1990b) Nomenclature of platelet-specific antigens. Hum Immunol 29:1–2
Yun G, Tolar J, Yerich AK, Marsh SGE, Robinson J, Noreen H, Blazar BR, Miller JS (2007) A novel method for KIR-ligand typing by pyrosequencing to predict NK cell alloreactivity. Clin Immunol 123:272–280

The IMGT/HLA Database

James Robinson and Steven G.E. Marsh

Introduction

Background

The IMGT/HLA Database was launched to provide a specialist database for the allelic sequences of the genes in the HLA system, also known as the human Major Histocompatibility Complex (MHC). This complex of over 4 megabases is located within the 6p21.3 region of the short arm of human chromosome 6 and contains in excess of 220 genes (Horton et al. 2004). The core genes of interest in the HLA system are 21 highly polymorphic HLA genes, that influence the outcome of cell and organ transplants and mediate the host response to infectious disease. Nucleotide sequences for nearly 3,000 different alleles of these genes have been determined. HLA genes are divided into class I (HLA-A, -B and -C) or class II (HLA-DR, -DQ, -DP) genes depending on the structure and function of their protein products. The naming of new HLA genes and allele sequences and their quality control is the responsibility of the WHO Nomenclature Committee for Factors of the HLA System (Bodmer et al. 1999; Marsh et al. 2001).

A Historical Perspective

The sequencing of HLA alleles first began in the late 1970s using protein sequencing techniques predominately to determine HLA class I alleles – the first complete HLA class I allele sequence, B7.2 now known as B*070201, being published in 1979 (Orr et al. 1979). It was a few years later in 1982 that the first HLA class II allele, a DRA

S.G.E. Marsh (✉)
The Anthony Nolan Research Institute, Royal Free Hospital, Pond Street, Hampstead, NW3 2QG, London, UK
UCL Cancer Institute, Royal Free Campus, Pond Street, Hampstead, NW3 2QG, London, UK
e-mail: marsh@ebi.ac.uk

D.R. Flower et al. (eds.), *Bioinformatics for Immunomics*, Immunomics Reviews 3, DOI 10.1007/978-1-4419-0540-6_3, © Springer Science+Business Media, LLC 2010

allele, was determined by more conventional cDNA sequencing (Lee et al. 1982). In 1987 the first HLA DNA sequences or alleles were named by the WHO Nomenclature Committee for Factors of the HLA System (Bodmer et al. 1989). Previous to this only the serologically defined antigens had been given official designations. At this time some 12 class I alleles were named A*0201–0204, B*0701–0702 and B*2701–2706, together with nine class II alleles, DRB1*0401–0405, DRB3*0101, 0201, 0301 and DRB4*0101. Although many other alleles had already been defined the committee did not consider them at that time. Two years later, in 1989, the Nomenclature Committee met for the first time outside the auspices of an International Histocompatibility Workshop to assign official allele names to the large number of HLA allele sequences that were by that time being published regularly. A total of 56 novel class I alleles and 78 class II alleles were named (Bodmer et al. 1990). It soon became apparent that the analysis and assigning of official names to alleles could not wait for either periodic histocompatibility workshops or even annual Nomenclature Committee meetings, and so began the process of assessing newly defined HLA allele sequences. Julia Bodmer and Steven Marsh at the Imperial Cancer Research Fund (ICRF) in collaboration with Peter Parham at Stanford University carried out this work. It was out of the need to record and manage the HLA sequence data being submitted to the Nomenclature Committee that came the first incarnation of an HLA Sequence Databank (HLA-DB) (Marsh and Bodmer 1993). Periodically HLA class I (Zemmour and Parham 1991, 1992; Arnett and Parham 1995; Mason and Parham 1998) and class II (Marsh and Bodmer 1990, 1991, 1992, 1994, 1995; Marsh 1998) sequence alignments were published in a variety of journals, and by 1995 the numbers of new alleles being reported warranted the publication of monthly nomenclature updates (Marsh 2008), something which continues to this day, Fig. 1. By 1995, the expansion of the Internet and the introduction of the World Wide Web (WWW) saw the first distribution of the HLA sequence alignments from the web pages of the Tissue Antigen Laboratory at the ICRF. This work was transferred to the Anthony Nolan Research Institute (ANRI) in 1996 where it continues to this day. The IMGT/HLA Database (Robinson et al. 2000, 2001, 2003; Robinson and Marsh 2000) began in 1997 as part of a European collaboration involving the ICRF, ANRI and the European Bioinformatics Institute (EBI) who maintain the European Molecular Biology Laboratory's nucleotide sequence database (EMBL). The work was initially funded by grants from the European Union, BIOMED1 (BIOCT930038) and BIOTECH2 (BIO4CT960037), awarded to the ICRF as part of the International ImMunoGeneTics (IMGT) database project (Giudicelli et al. 1997). The IMGT/HLA database was first released in 1998; the database combines the sequence data and information previously provided to the WHO Nomenclature Committee for Factors of the HLA System and the additional data found in the original EMBL/GenBank/DDBJ entries.

The Database Today

The database contains entries for all HLA alleles officially named by the WHO Nomenclature Committee for Factors of the HLA System. These entries are derived

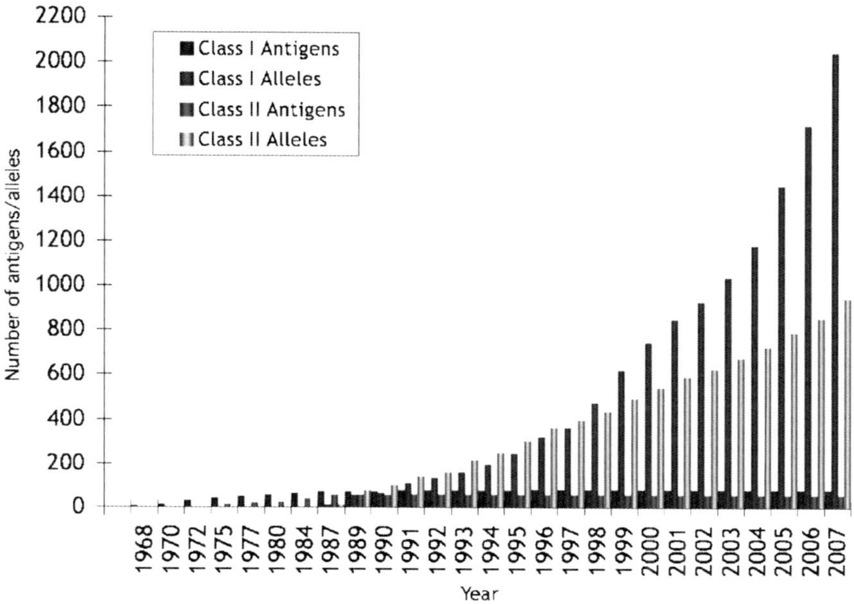

Fig. 1 Numbers of HLA Class I and II alleles officially recognised by the WHO Nomenclature Committee for Factors of the HLA System 1987 – 2007.

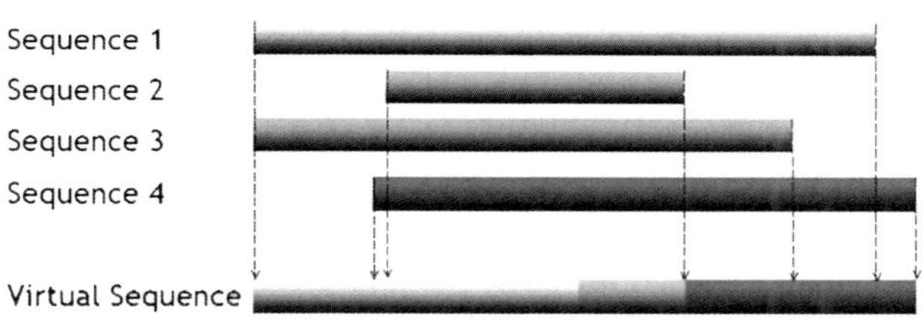

Fig. 2 Construction of a virtual sequence. A virtual sequence is constructed for each allele, representing the longest available sequence for each allele. This is generated by combining the component entries from EMBL/GenBank/DDBJ to create a single entry. The diagram shows four component entries aligned and merged to form a virtual sequence

from expertly annotated copies of the original EMBL/GenBank/DDBJ entries. This means that the IMGT/HLA database can contain a number of different entries for any single allele. In order to store all the information for each allele in a single entry available to the user, the virtual sequence concept was developed, Fig. 2. In the database a virtual sequence represents the combination of all the EMBL/GenBank/DDBJ entries for a single allele combining to form a single expertly annotated entry. These component entries are submitted to the database in the form of IMGT/HLA submissions either by the original author or by our curators when sequences of interest have been identified by data mining but have yet to be submitted to the database. The nucleotide sequences are then aligned using the ClustalX program (Thompson et al. 1994). This provides a multiple sequence alignment of the component sequence using a recognised alignment program. The alignment produced is then reduced to the single longest contiguous sequence for the submitted allele. Previous published alignments sometimes used a consensus sequence for alignments; however, following guidelines issued by the Human Gene Nomenclature Committee (Antonarakis 1998), the IMGT/HLA database instead uses a reference allele sequence at each locus. Insertions, indicated by periods (.), are added to the virtual sequence to ensure alignment to the reference sequence. To distinguish the IMGT/HLA entries from the component EMBL entries each new allele is assigned a unique accession number. The accession numbers follows the format HLA00000, where the '00000' represents a numerical code. The EMBL accession numbers are not used as primary identifiers in IMGT/HLA because many alleles are derived from multiple EMBL sequence entries.

Accessing the Database

The first public release of the IMGT/HLA Database was made on 16 December 1998 and was included on the EBI web server as part of the IMGT project. The database is updated every 3 months to include all the publicly available sequences officially named by the WHO Nomenclature Committee since the last release of the database. With each release all the tools are updated to include the new sequences, and information on all the new and modified sequences is reported. The previous release is archived for reference.

The main access point for the user is the World Wide Web, which allows the users to employ a number of search tools and other facilities to retrieve, manipulate and analyse HLA data. This is all done through the custom written Common Gateway Interface (CGI) scripts available at the IMGT/HLA website. The IMGT/HLA website can be split into three main areas. First, information and help pages that provide background on the database, provide in-depth help on the tools and data available and documentation of the IMGT/HLA file formats. The second area includes the tools designed specifically for the IMGT/HLA database. These allow the user to perform sequence analysis and retrieval. The final pages are links to sequence-analysis tools at the EBI, including SRS, BLAST and FASTA. The tools available from the website will be discussed in detail later. The core tools allow the users to perform sequence alignments, allele queries and sequence searches. This is done by combining the custom-built tools, with existing tools already available from the EBI such as BLAST, FASTA and SRS. Access to all the pages and tools is via the IMGT/HLA Database homepage, see Fig. 3. Other access points for the user include the File Transfer Protocol (FTP) server on the EBI Web Site; in addition a subset of the data stored in the IMGT/HLA database is also provided in a text-based format at the HLA Informatics Group web pages of the Anthony Nolan Research Institute website.

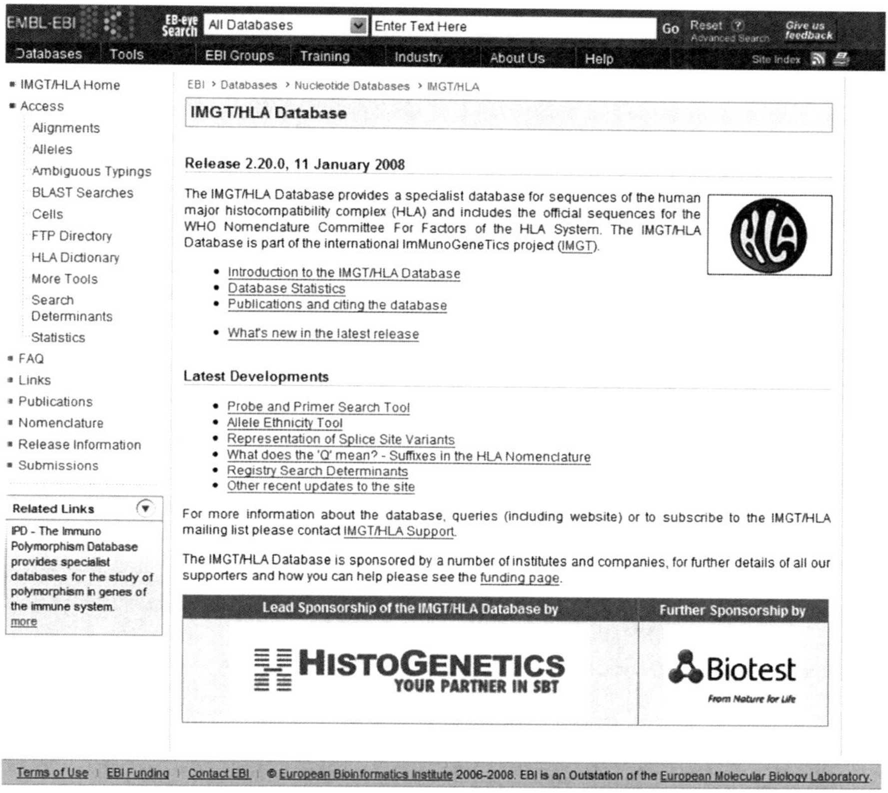

Fig. 3 The IMGT/HLA Database Homepage, the main access point to the online tools provided by the database

IMGT/HLA Tools

Allele Search Tools

The most used tool on the IMGT/HLA website is the 'Allele Query Tool'. This is designed to allow the user to retrieve the full sequence and information pertaining to any officially named allele. The tool is available directly from the homepage or from the tools section of the website. The tool provides a simple to use interface for retrieving allele information. Similar searches can be performed using the SRS interface (discussed later), but this requires more knowledge of the tools, file structure and data. The 'Allele Query Tool' requires only a search term in order to retrieve a report on any allele in the database. The search tool allows different resolutions of the allele's numeric code to allow flexibility in searches. For example the A*0101 designation could refer to one of six alleles, A*01010101,

A*01010102N, A*010102, A*010103, A*010104 and A*010105, and entering 'A*0101' will retrieve all these alleles. Working back from this entering just 'A*01', would retrieve the 35 alleles with the 'A*01' designation. This facility can be exploited to gain a list of all alleles at any locus by entering a '*locus**', e.g. 'B*' for HLA-B, in the box to retrieve the full list. Other shortcuts include the retrieval of null alleles for a specific locus; by entering '*locus**N' into the box you can retrieve a list of all non-expressed or null alleles. The output of the search lists all the allele names that correspond to the search term provided; these then provide a hypertext link to the full entry for each allele. The output provided for each allele includes the official allele designation, previously used designations and the unique IMGT/HLA accession number, which is a link to the IMGT/HLA flat file. Other information provided includes the date that the allele was named and its current status (as some allele designations have been deleted) and information on the individual or cell line from which the sequence was derived. Links to all component EMBL/GenBank/DDBJ entries are also included. Recently information from the HLA Dictionary (Schreuder et al. 2005) has also been added to some entries. The dictionary presents the serological equivalents of HLA-A, -B, -C, -DRB1, -DRB3, -DRB4, -DRB5 and -DQB1 alleles. The data summarise equivalents obtained by the WHO Nomenclature Committee for Factors of the HLA System, the International Cell Exchange (UCLA), the National Marrow Donor Program (NMDP), the 13th International Histocompatibility Workshop, recent publications and individual laboratories. Any published references are also included with, wherever possible, a link to the PubMed entry for that citation. The PubMed link provides an online version of the abstract as well as links to other citations by the author and to similar papers. The final section of the output details the official nucleotide and protein sequence as well as where available any genomic sequence for that allele.

The IMGT/HLA database is also available in a flat file format, which is standard format for text files. These flat files follow the EMBL flat file format and provide a standard release format. The flat files utilise the unique accession number and assign a standardised description and keywords to all entries. This accession number is used in the EBI tools to link back to the flat file entry. The flat files also contain the first release of the IMGT/HLA features. The sequence features currently used are a small subset of the standard set used by EMBL, but as the database continues to develop further features may be added. The initial feature qualifiers cover the source (cell of origin), the coding sequence (cds), exon boundaries and the protein translation. Other information provided by the flat file replicates the allele query tool output. The only additional information included is a list of all the component sequence entries and other cross-references to sequence databases like SWISSPROT, TREMBL and PDB. These links are to the original entries and so the files retrieved may differ from the IMGT/HLA entry due to the annotation procedure. These flat files can be searched using Sequence Retrieval System (SRS) tool (Etzold et al. 1996) which is an advanced search tool for interrogating flat files, and the IMGT/HLA flat files are included in the EBI SRS libraries. These interrogations can range from very simple queries, such as searching for an accession number or keyword, to more complex searches such as for authors of papers describing an allele. In order to use SRS some familiarity with the flat file format is required. This tool allows the user to search on any of the sequence features, the accession numbers, keywords assigned to the sequences or the description, and is probably the best method of retrieving HLA sequences from a general nucleotide database. Once users are accustomed to the way data are presented they can quickly build up very complex queries. Another advantage of the SRS tool is that it can also be used to launch other applications, e.g. BLAST, Clustal, on any query results. Therefore it is possible to retrieve all the flat files for a certain subset of data and automatically load these into the Clustal alignment tool for example. The SRS tool also allows the users to customise the output of searches, so that they can quickly see how relevant entries are to your search criteria. Tutorials for the SRS search engine are available from the EBI and SRS website. SRS can be found at the EBI website and can be used to search a number of different databases. The information in the IMGT/HLA entry should therefore be taken as the definitive source in cases of disagreement.

The IMGT/HLA Database is also involved in developing data format standards for HLA information exchange. This work in collaboration with other immunoinformatics groups will provide both an XML output format for the IMGT/HLA database and an XML reporting format for tissue typing laboratories. The XML output will contain similar information to that described for the flat files and allele output.

Searching for Similar Sequences

The first type of search that many people do is to look for a particular allele by name or by a certain characteristic such as the cell or author. The main alternative to this is to search on the actual sequence and not on the name or keywords. Sequence similarity searches look for sequence matches in a query sequence against a reference database of known sequences. The accuracy of these matches is based on a number of similarity measures, and in general retrieve identical or highly similar sequences. The IMGT/HLA database is included as a library for searching within the EBI's Similarity & Homology toolset. These include well-known tools like BLAST & FASTA. BLAST stands for Basic Local Alignment Search Tool and it can be used to compare a sequence with those contained in nucleotide and protein databases by aligning the novel sequence with previously characterised genes. The emphasis of this tool is to find regions of sequence similarity, which will yield functional and evolutionary clues about the structure and function of this novel sequence. FASTA stands for FAST-All and can be used for a fast protein comparison or a fast nucleotide comparison. This program achieves a high level of sensitivity for similarity searching at high speed. Both these tools can be used for searching libraries of HLA nucleotide and protein sequences.

Viewing Alignments of HLA Sequences

Another popular use of the IMGT/HLA database is to view multiple sequence alignments. HLA allele sequences may differ from each other by as little a single nucleotide, over a genomic sequence of 3,300 bases. These alignments allow a visual interpretation of sequence similarity, so that polymorphic positions can easily be identified and motifs found in multiple alleles are easily identified (Fig. 4). The representation of HLA sequences in this manner can be useful when designing reagents for HLA typing, such as primers or oligonucleotide probes. The sequence alignments are available via a link from the main IMGT/HLA homepage; they can also be found using the tools page of the website. The interface provided lets the

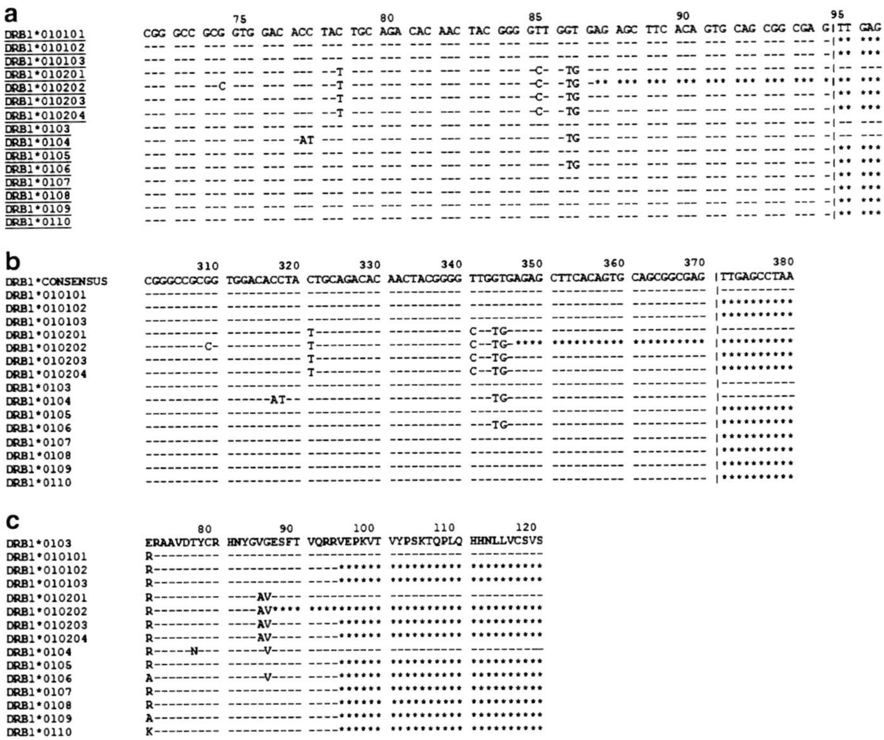

Fig. 4 Alignment formats available from the IMGT/HLA Database. The examples shown are based on alignment **a** which shows 15 DRB1*01 alleles. In these alignments a *dash* indicates identity to the reference sequence and an *asterisk* denotes an unsequenced base. In alignment **a** the allele names are *underlined*, as they are hyperlinks to the allele's entry, and the nucleotide sequence is displayed in codons. Alignment **b** shows how an alternative reference sequence can be used; here, for example we have used a DRB1 consensus sequence. The sequence is displayed in blocks of ten nucleotides. Alignment **c** represents a translation of the nucleotide sequence to produce a protein sequence alignment. In this final example the DRB1*0103 allele has been used as the reference sequence

user define a number of key variables for the alignments, before producing an online output, which can be printed or downloaded. The first step in any alignment is to select the locus of interest. The tool provides a drop-down list of all loci available and also includes some additional options like class I (all HLA-A, HLA-B and HLA-C alleles) and different HLA-DRB gene combinations. The selection of a locus automatically updates the list of features which can be aligned, as well as the default reference sequence used for the alignment. The types of feature available for alignment are the nucleotide coding sequence and individual exons, the genomic sequences and individual introns (where available), the signal peptide, mature protein and full-length protein sequence. In addition there are some commonly requested regions like a single alignment of both exons 2 and 3 or exons 2, 3 and 4 which have been included to aid in the analysis of sequence-based typing results. Genomic sequence is currently only available for the HLA class I loci; however, work is underway on class II genomic sequences for inclusion into the alignment tool. The alignment tool options also allow the user to display a subset of alleles of a particular locus, omit alleles unsequenced for a particular region and align against a particular reference or consensus sequence. The alignment tool uses standard formatting conventions for the display of sequence alignments. The alignment tool does not perform a sequence alignment each time it is used, but it extracts pre-aligned sequences, allowing for faster access.

The alignments adhere to a number of conventions for displaying evolutionary events and numbering. The numbering of the alignments is based upon the sequence of the reference allele. For a nucleotide sequence, the A of the initiation Methionine codon is denoted nucleotide +1 and the nucleotide 5′ to +1 is numbered −1. There is no nucleotide zero (0). All numbering is based on the ATG of the reference sequence. If a nucleotide sequence is displayed in codons, then the protein numbering is applied.

For amino acid-based alignments, the first codon of the mature protein, after cleavage of the signal sequence, is labelled codon 1, and the codon 5′ to this is numbered −1. In all sequences the following conventions are used. Where identity to the reference sequence is present the base will be displayed as a hyphen (-). Non-identity to the reference sequence is shown by displaying the appropriate base at that position. Where an insertion or deletion has occurred this is represented by a period (.). If the sequence is unknown at any point in the alignment, this is be represented by an asterisk (*). In protein alignments for null alleles, the 'Stop' codons are represented by an X and further sequence following the termination codon is not displayed and appears blank. The alignment tool also marks up certain parts of a sequence when splice site variants are present. In a number of alleles there are mutations very close to the exon–intron splice site. When these occur the splicing may not occur as predicted. In order to maintain alignment with other alleles, the alternatively spliced sequence is shaded to identify the effect of the mutation on the CDS.

The flexibility of the new alignment tool means that unlike previous alignments you can now display a small subset of sequences against an allele of your choice, using a number of display options. The previous text alignments are still requested and as a result are available from the ANRI website and in a zipped file in the FTP directory. Users are also able to use locally available software to produce their own alignments as the FTP directory contains files in popular formats for them to download and import.

Submitting New Sequences

As well as providing HLA sequences for retrieval the IMGT/HLA website also provides the tools for submitting both new and confirmatory sequences to the WHO HLA Nomenclature Committee. This is now the only accepted method for submitting sequences to the Committee, who strongly discourage the use of numbers or names for alleles, genes or specificities which pre-empt assignment of official designations. Submissions are processed, which incorporates automated analysis and annotation of the sequence, and then given an official name, before being loaded into the IMGT/HLA database and included in the monthly nomenclature reports (Marsh 2005). The submission tool can be used for both new and confirmatory sequences, and is capable of holding confidential entries until a set time, thus allowing alleles to be named before publication. The submission of new HLA sequences to the IMGT/HLA database does not replace the submission of these sequences to EMBL/GenBank/DDBJ, as the submission criteria state that the sequences must also have been submitted to these databases.

Recent Developments

Recent developments to the website have seen the addition of search tool for identifying primer and probe sequences. Many HLA typing laboratories have spreadsheets detailing probe hit patterns for different alleles. Each time a new database was released it was necessary to update these ever-expanding lists. The new Probe & Primer Search Tool allows users to enter a list of primer sequences, and the tool will search the known coding sequences for these and report any matches in a file format suitable for cutting and pasting into existing spreadsheets. The tool is currently limited to coding sequence, but as the number of genomic sequences in the database expands the tool can be modified to search these regions as well.

HLA Sequences in the Generalist Databanks

It should be noted that all sequences within the IMGT/HLA database should also be available from the more general nucleotide sequence databases, the EMBL

(Kanz et al. 2005; Cochrane et al. 2008), GenBank (Benson et al. 2008) and the DNA Database of Japan (DDBJ) (Sugawara et al. 2008; Tateno 2008). The generalist databanks are not HLA specific, but rather large international data repositories of gene sequences from for all organisms. These three databases form an international collaboration and exchange sequences daily so that each contains identical data. Most published sequences can be found in these databases. The advantages of using the large generalist sequence databases are the large number of sequences available, covering a wide range of data relating to HLA. The main problem when accessing HLA sequences from these databases lies in the definition of the sequence. Upon submission a number of steps are taken to ensure that the sequence is accurately described and up to date. The author then assigns keywords and description, and these can vary. Despite the work of the WHO Nomenclature Committee for Factors of the HLA System in monitoring HLA allele designations and maintaining the sequences, they have no control of how sequences are defined in these generalist databases. Readers should therefore be aware that entries may be incorrectly named, contain unofficial designations or contain sequencing errors.

Conclusions

IMGT/HLA database provides a centralised resource for everybody interested, clinically or scientifically, in the HLA system. The database and accompanying tools allow the study of all HLA alleles from a single site on the World Wide Web. It should aid in the management and continual expansion of HLA nomenclature, providing an ongoing resource for the WHO Nomenclature Committee.

Acknowledgements We would like to thanks Matthew Waller and Sylvie Fail for their work on development and curation of the database. We would like to acknowledge the support of the following organisations: Histogenetics, Abbott Laboratories Inc., the American Society for Histocompatibility and Immunogenetics , the Anthony Nolan Bone Marrow Trust, Biotest, BAG Healthcare, Dynal, the European Federation for Immunogenetics, Qiagen, One Lambda Inc, Invitrogen, Innogenetics, the National Marrow Donor Program, the Marrow Foundation and Tepnel Lifecodes. Initial support for the IMGT/HLA database project was from the Imperial Cancer Research Fund (now Cancer Research UK) and an EU Biotech grant (BIO4CT960037).

Appendix : Access and Contact

IMGT/HLA Homepage	http://www.ebi.ac.uk/imgt/hla/
IMGT/HLA FTP Site	ftp://ftp.ebi.ac.uk/pub/databases/imgt/mhc/hla/
Contact	hladb@ebi.ac.uk

References

Antonarakis SE (1998) Recommendations for a nomenclature system for human gene mutations. Nomenclature Working Group. Hum Mutat 11:1–3

Arnett KL, Parham P (1995) HLA class I nucleotide sequences, 1995. Tissue Antigens 46:217–257

Benson DA, Karsch-Mizrachi I, Lipman DJ, Ostell J, Wheeler DL (2008) GenBank. Nucleic Acids Res 36:D25–D30

Bodmer WF, Albert E, Bodmer JG, Dupont B, Mach B, Mayr WR, Sasazuki T, Schreuder GMT, Svejgaard A, Terasaki PI (1989) In: Dupont B (ed) Immunobiology of HLA, vol. 1, Springer, New York, pp 72–79

Bodmer JG, Marsh SGE, Parham P, Erlich HA, Albert E, Bodmer WF, Dupont B, Mach B, Mayr WR, Sasasuki T, Schreuder GMT, Strominger JL, Svejgaard A, Terasaki PI (1990) Nomenclature for factors of the HLA system, 1989. Tissue Antigens 35:1–8

Bodmer JG, Marsh SGE, Albert ED, Bodmer WF, Bontrop RE, Dupont B, Erlich HA, Hansen JA, Mach B, Mayr WR, Parham P, Petersdorf EW, Sasazuki T, Schreuder GM, Strominger JL, Svejgaard A, Terasaki PI (1999) Nomenclature for factors of the HLA system, 1998. Tissue Antigens 53:407–446

Cochrane G, Akhtar R, Aldebert P, Althorpe N, Baldwin A, Bates K, Bhattacharyya S, Bonfield J, Bower L, Browne P, Castro M, Cox T, Demiralp F, Eberhardt R, Faruque N, Hoad G, Jang M, Kulikova T, Labarga A, Leinonen R, Leonard S, Lin Q, Lopez R, Lorenc D, McWilliam H, Mukherjee G, Nardone F, Plaister S, Robinson S, Sobhany S, Vaughan R, Wu D, Zhu W, Apweiler R, Hubbard T, Birney E (2008) Priorities for nucleotide trace sequence and annotation data capture at the Ensembl Trace Archive and the EMBL Nucleotide Sequence Database. Nucleic Acids Res 36:D5–D12

Etzold T, Ulyanov A, Argos P (1996) SRS: information retrieval system for molecular biology data banks. Methods Enzymol 266:114–128

Giudicelli V, Chaume D, Bodmer J, Muller W, Busin C, Marsh SGE, Bontrop R, Marc L, Malik A, Lefranc MP (1997) IMGT the international ImMunoGeneTics database. Nucleic Acids Res 25:206–211

Horton R, Wilming L, Rand V, Lovering RC, Bruford EA, Khodiyar VK, Lush MJ, Povey S, Talbot CC Jr, Wright MW, Wain HM, Trowsdale J, Ziegler A, Beck S (2004) Gene map of the extended human MHC. Nat Rev Genet 5:889–899

Kanz C, Aldebert P, Althorpe N, Baker W, Baldwin A, Bates K, Browne P, van den Broek A, Castro M, Cochrane G, Duggan K, Eberhardt R, Faruque N, Gamble J, Diez FG, Harte N, Kulikova T, Lin Q, Lombard V, Lopez R, Mancuso R, McHale M, Nardone F, Silventoinen V, Sobhany S, Stoehr P, Tuli MA, Tzouvara K, Vaughan R, Wu D, Zhu W, Apweiler R (2005) The EMBL Nucleotide Sequence Database. Nucleic Acids Res 33:D29–D33

Lee JS, Trowsdale J, Travers PJ, Carey J, Grosveld F, Jenkins J, Bodmer WF (1982) Sequence of an HLA-DR alpha-chain cDNA clone and intron-exon organization of the corresponding gene. Nature 299:750–752

Marsh SGE (1998) HLA class II region sequences, 1998. Tissue Antigens 51:467–507

Marsh SGE (2005) Nomenclature for factors of the HLA system, update September 2005. Tissue Antigens 67:94–95

Marsh SGE (2008) Nomenclature for factors of the HLA system, update December 2007. Tissue Antigens 71:262–263

Marsh SGE, Bodmer JG (1990) HLA-DRB nucleotide sequences, 1990. Immunogenetics 31:141–144

Marsh SGE, Bodmer JG (1991) HLA class II nucleotide sequences, 1991. Tissue Antigens 37:181–189

Marsh SGE, Bodmer JG (1992) HLA class II nucleotide sequences, 1992. Tissue Antigens 40:229–243

Marsh SGE, Bodmer JG (1993) HLA Class II Sequence Databank. Hum Immunol 36:44

Marsh SGE, Bodmer JG (1994) HLA class II region nucleotide sequences, 1994. Eur J Immunogenet 21:519–551

Marsh SGE, Bodmer JG (1995) HLA class II region nucleotide sequences, 1995. Tissue Antigens 46:258–280

Marsh SGE, Bodmer JG, Albert ED, Bodmer WF, Bontrop RE, Dupont B, Erlich HA, Hansen JA, Mach B, Mayr WR, Parham P, Petersdorf EW, Sasazuki T, Schreuder GM, Strominger JL, Svejgaard A, Terasaki PI (2001) Nomenclature for factors of the HLA system, 2000. Tissue Antigens 57:236–283

Mason PM, Parham P (1998) HLA class I region sequences, 1998. Tissue Antigens 51:417–466

Orr HT, Lopez de Castro JA, Lancet D, Strominger JL (1979) Complete amino acid sequence of a papain-solubilized human histocompatibility antigen, HLA-B7. 2. Sequence determination and search for homologies. Biochemistry 18:5711–5720

Robinson J, Marsh SGE (2000) The IMGT/HLA sequence database. Rev Immunogenet 2:518–531

Robinson J, Malik A, Parham P, Bodmer JG, Marsh SGE (2000) IMGT/HLA database – a sequence database for the human major histocompatibility complex. Tissue Antigens 55:280–287

Robinson J, Waller MJ, Parham P, Bodmer JG, Marsh SGE (2001) IMGT/HLA Databasem – a sequence database for the human major histocompatibility complex. Nucleic Acids Res 29:210–213

Robinson J, Waller MJ, Parham P, de Groot N, Bontrop R, Kennedy LJ, Stoehr P, Marsh SGE (2003) IMGT/HLA and IMGT/MHC: sequence databases for the study of the major histocompatibility complex. Nucleic Acids Res 31:311–314

Schreuder GM, Hurley CK, Marsh SGE, Lau M, Fernandez-Vina M, Noreen HJ, Setterholm M, Maiers M (2005) The HLA Dictionary 2004: a summary of HLA-A -B, -C, -DRB1/3/4/5 and -DQB1 alleles and their association with serologically defined HLA-A,-B, -C, -DR and -DQ antigens. Int J Immunogenet 32:19–69

Sugawara H, Ogasawara O, Okubo K, Gojobori T, Tateno Y (2008) DDBJ with new system and face. Nucleic Acids Res 36:D22–D24

Tateno Y (2008) [International collaboration among DDBJ EMBL Bank and GenBank]. Tanpakushitsu Kakusan Koso 53:182–189

Thompson JD, Higgins DG, Gibson TJ (1994) Improved sensitivity of profile searches through the use of sequence weights and gap excision. Comput Appl Biosci 10:19–29

Zemmour J, Parham P (1991) HLA class I nucleotide sequences, 1991. Tissue Antigens 37:174–180

Zemmour J, Parham P (1992) HLA class I nucleotide sequences, 1992. Tissue Antigens 40:221–228

Ontology Development for the Immune Epitope Database

Jason A. Greenbaum, Randi Vita, Laura M. Zarebski, Alessandro Sette, and Bjoern Peters

A key challenge in bioinformatics today is ensuring that biological data can be unequivocally communicated between experimentalists and bioinformaticians. Enabling such communication is not trivial, as every scientific field develops its own jargon with implicit understandings that can easily escape an outsider. We describe here our approach to enforce an explicit and exact data representation for the Immune Epitope Database (IEDB Peters et al. 2005) through the use of a formal ontology.

Being the first database of its scale in the immune epitope domain, it was necessary for the IEDB to devise an adequate data structure at the outset of the project with the goal that it should be capable of capturing the context of immune recognition. Early on, it became readily apparent that an unambiguous description of the information being captured is imperative for consistent curation across journal articles and among curators. Accordingly, an initial ontology was developed (Sathiamurthy et al. 2005) based upon consultations with domain experts and guidance from expert ontologists. The structure devised from this ontology proved capable of dealing with a great deal of immunological data over time.

However, after several years of curation, it became necessary to adjust the ontology and data structure to accommodate more and more exceptions. Moreover, the database had, from the beginning, excluded particular types of experiments as they were not easily accommodated in its structure. These experiments, known as adoptive transfer or passive immunization, are complex in nature, involving one organism in which the immune response is generated (donor organism) and another to which that immune response is transferred and studied (recipient organism). In order to accommodate adoptive transfer experiments, as well as improve how all experimental approaches were represented, the IEDB has recently developed a new version of its ontology (ONTology of Immune Epitopes, ONTIE 2.0) and simultaneously restructured its database scheme.

This book chapter gives an introduction into the biology of immune epitope recognition followed by a brief primer on ontology development in general, and

B. Peters (✉)
La Jolla Institute for Allergy and Immunology, 9420 Athena Circle La Jolla, CA, 92037, USA
e-mail: bpeters@liai.org

D.R. Flower et al. (eds.), *Bioinformatics for Immunomics*, Immunomics Reviews 3,
DOI 10.1007/978-1-4419-0540-6_4,

introduces the higher level ontologies we are building upon. We then present results from our specific approach to develop ONTIE 2.0, and the global reorganization of the IEDB data structure.

Background: Immune Epitopes and the IEDB

The vertebrate immune system has the capacity to detect nonself molecules through adaptive immune receptors. These receptors include the B cell receptors (BCRs), present on the surface of B cells and secreted as antibodies, and the T cell receptors (TCRs). The molecular entities recognized by BCRs and TCRs are referred to as immune epitopes. Epitopes found in proteins consist of either short linear stretches of amino acids (i.e., peptides) or are conformational epitopes, formed by the spatial arrangement of several amino acid residues within the three-dimensional protein structure. Epitopes can also be derived from other molecules, including carbohydrates and lipids.

BCRs and antibodies bind directly to their targets, tagging them for further action by immune effector cells, and sometimes interfering directly with the antigen's function. TCRs recognize their target epitopes as part of a complex with a major histocompatibility complex (MHC) molecule. MHC molecules are found on the surface of antigen presenting cells and display a sample of peptides derived from digested proteins present inside the cell. After infection with a foreign pathogen, some MHC molecules will present peptides derived from nonself proteins. These peptide-MHC complexes are scanned by the TCR, allowing the T cell to probe the contents of cells and take action when nonself proteins are encountered.

The IEDB was created with the aim to gather all published experimental data characterizing immune epitopes. Details of the experimental context in which the epitopes are defined are captured along with the epitope structure itself. Capturing the experimental context is imperative, since being an epitope is not an intrinsic property of a molecular entity. In other words, to accurately describe an epitope requires not just detailing the molecular structure alone, but it must include a description of the context in which that molecular structure is recognized as an epitope by the immune receptors.

The data required to capture the experimental context are quite complex. The experiments performed to define epitopes are varied and comprise multiple parts and steps. For example, the first step of the experiment shown in Fig. 3 will require capturing:

(1) The host organism in which the immune response was studied (C57BL/6 mouse);
(2) The immunization processes that primed the immune response (subcutaneous injection); and
(3) The immunogen used (SARS coronavirus nucleoprotein).

By providing the details of the experimental context, comparisons across different studies can be made and discrepancies between experiments can be better understood.

Additionally, refined retrieval of specific subsets of data (e.g., epitopes defined in humans, or those recognized by CD4+ T cells) is also made possible.

To capture the experimental context in which epitopes are defined, the IEDB utilizes over 300 database fields to describe an individual experiment. The database is primarily populated from experiments described in articles published in peer-reviewed, PubMed-indexed journals. The translation of the published data into the database fields is performed by a team of nine Ph.D.-level curators following a formalized curation process, with several quality control steps and mechanisms to enforce and adapt curation rules (Salimi and Vita 2006; Vita et al. 2006). Currently, more than 5,000 articles have been curated for a total of greater than 115,000 assays involving approximately 38,500 epitopes. The IEDB ontology provides the formal framework to enforce consistency of curation for this vast amount of data.

Background: Why Ontology Development?

Ontology is a fundamental branch of philosophy that goes back to the early Greek philosophers Aristotle and Plato and deals with the question of "what things exist?" This question is answered by listing individual entities, types of entities, and relationships among them.

The use of ontologies to annotate data in biological databases has become ubiquitous in recent years (Ashburner et al. 2000; Rosse and Mejino 2003; Whetzel et al. 2006a, b). This is a result of their increased application in the information sciences in general. Information science has revolutionized the ability to retrieve specific sets of records or to derive summary statistics on data by providing the capacity to store large sets of information in databases. To be useful, the meaning of the stored information has to be clear to all users. Agreement between the users of the database and the data providers regarding how each entry in each field is interpreted can be achieved by an unambiguous mapping to a formal ontology.

A formal ontology represents the most complex of three approaches commonly used in information science to enforce data consistency:

- *Controlled vocabulary* – a list of terms and their definitions
- *Taxonomy* – the terms of a controlled vocabulary are organized into a hierarchical structure, typically an *is a* hierarchy (parent – child)
- *Formal ontology* – the *is a* hierarchy of a taxonomy is expanded to include multiple types of relationships between terms, such as *part of* and *source of*

Fig. 1 illustrates how several terms from an immunization protocol (Fig. 3) would be represented in each of these approaches. A controlled vocabulary (Fig. 1a) would arrange the terms as a list with each term having a specific definition, but cannot accommodate explicit relationships. Under a taxonomy (Fig. 1b), the terms are also explicitly defined and have exactly one *is a* relationship to their parent term. The fact that nucleoprotein *is a* protein and a protein *is a* object allows the transitive conclusion to be drawn that nucleoprotein *is a* object. Although this relationship

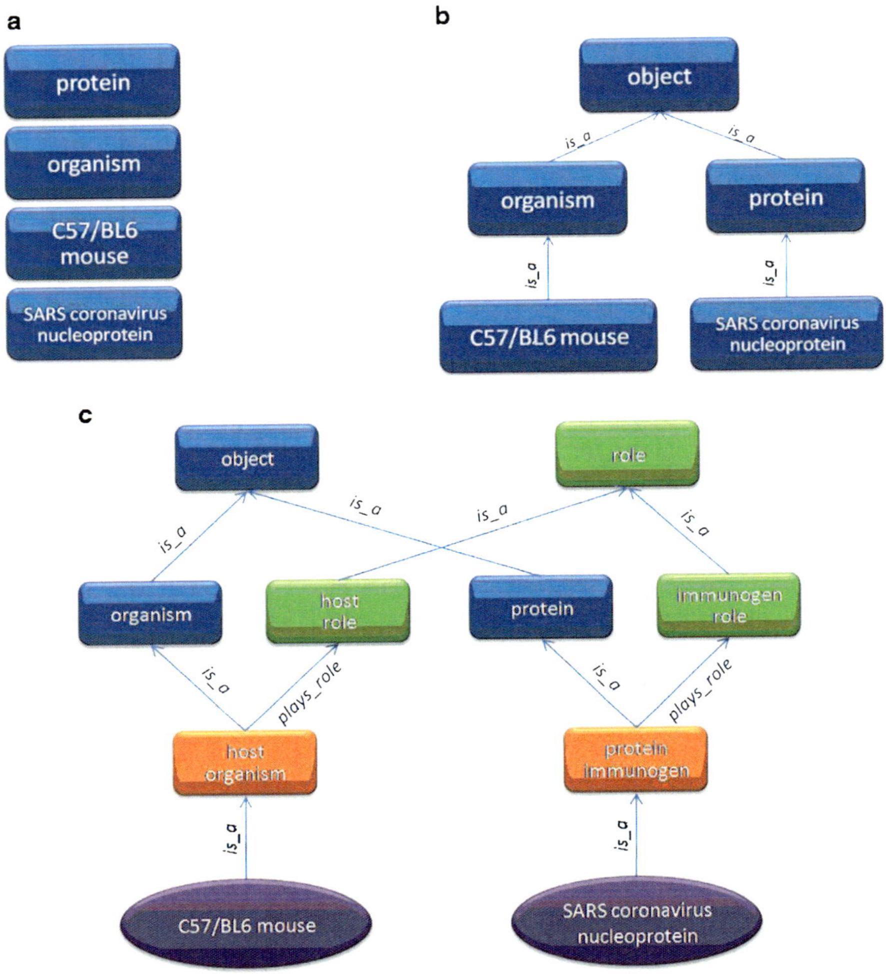

Fig. 1 Objects in an immunization assay represented in three different formats. (**a**) Controlled vocabulary – The terms are arranged as a list with definitions, but not with explicit relationships. (**b**) Taxonomy (Hierarchy) – The terms are arranged as an *is a* hierarchy, defining their basic classification. Each object has exactly one parent, except for the root node, which has none. (**c**) Formal ontology – In addition to the *is a* hierarchy the objects (blue rectangles) and roles (green rectangles) can have other relationships between them. For example, the protein immunogen is a derived class (orange rectangles) that *is a* protein that *plays role* immunogen. The SARS coronavirus nucleoprotein is an instance (purple ovals) of this class

may be obvious, the ability to express it in formal terms is a step towards automating the discovery of new relationships.

A formal ontology (Fig. 1c) expands upon the concept of a taxonomy in that it allows for multiple types of relationships between and among objects. Such additional relationships include *part of*, *derived from*, *plays role* and *proxy for*, and can be used to formalize the textual definitions of terms. For example, in the formal ontology depicted in Fig. 1c, the protein immunogen has two relationships; it *is a* protein and it *plays role* immunogen.

Standing on the shoulders of BFO, OBO, and OBI

One of the benefits of using a formal ontology for data annotation is the potential for re-use of existing ontologies upon which the domain-specific solution can be built. Although this could significantly reduce the amount of work to build the domain-specific ontology, the primary advantage of this approach is the enhanced interoperability between different resources that it enables.

For the 2.0 version of ONTIE, we have integrated our development work into the Ontology for Biomedical Investigations (OBI) effort (Whetzel et al. 2006a), in which we are actively participating. The scope of OBI is to provide the vocabulary necessary to describe any biomedical experiment, instrument, reagent and the like. By integrating the experimental terms needed for ONTIE into OBI, we are ensuring consistency and interoperability with the nearly 20 scientific communities included in the OBI effort.

OBI itself is a candidate for date for the open biomedical ontologies (OBO) foundry (Smith et al. 2007). While OBO is an umbrella group open to all ontology developers adhering to some format and availability requirements, the OBO foundry represents an effort to develop high quality, interoperable, and orthogonal ontologies. In short, many domain-specific ontologies are under development by the respective experts in their fields. The OBO foundry aims to enforce a set of standards to ensure that these ontologies will be interoperable and to avoid unnecessary duplication of work. Other candidate ontologies for the OBO foundry such as the Gene Ontology, Cell Ontology or Disease Ontology deal with the vocabulary necessary to describe different aspects of biological reality. This allows OBI to reference these other foundry ontologies to describe things like the cell types contained in a sample used in an experiment.

OBI is designed with the Basic Formal Ontology (BFO) (Grenon et al. 2004) as its upper level. BFO intends to be a high-level ontology from which more focused domain-specific ontologies can be developed. At its most basic level, BFO divides the world into continuants and occurrents. Continuants are entities that persist through time and are further divided into independent and dependent continuants. Independent continuants are defined as bearers of qualities and include OBI material entities (e.g., instruments, organisms, cells, proteins). Dependent continuants are entities that are intrinsic to or borne by other entities. These include things like qualities (e.g., mass, shape), and roles (drug, placebo) that require an independent continuant to exist. Occurrents make up the second major branch of the BFO hierarchy and are defined as entities that unfold over time. OBI further divides these into planned (e.g., experiments) and unplanned (e.g., diseases) processes. A partial schema of the hierarchy is depicted in Fig. 2.

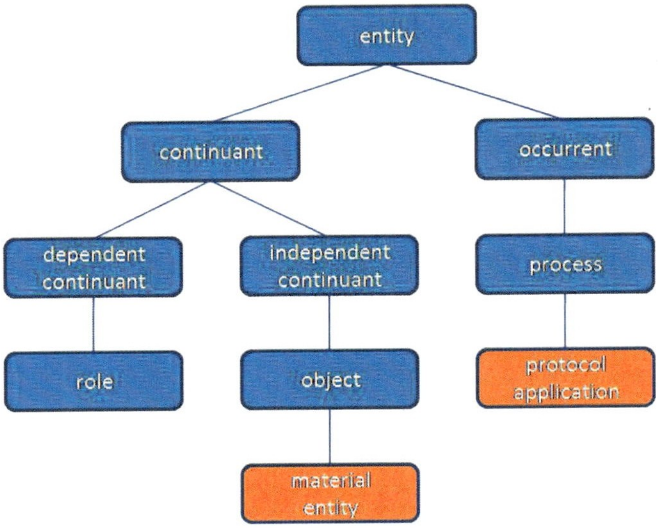

Fig. 2 *High Level BFO/OBI structure*. The high level BFO (*blue background*) and OBI (*orange background*) hierarchy is represented here. For simplification of the figure, only the most relevant classes are shown

In the next section, we describe how the integration into this high level structure has effected the development of ONTIE. It is important to point out that OBI and, to a lesser degree, BFO are still under active development and their topology may change. We are actively participating in this development and plan to keep ONTIE in sync with these ontologies.

Ontology Development for the IEDB

Objects, Roles, and Processes

The initial ontology for the IEDB (Sathiamurthy et al. 2005) was developed prior to population of the database, and was based upon consultations with domain experts and guidance from expert ontologists. It proved capable of dealing with the vast majority of epitope data encountered in the literature. However, it proved to be hard to extend to deal with novel types of data, contained internal inconsistencies and ambiguities that only became apparent over time, and provided only limited interoperability with other data resources. These experiences gained through several years of curation led to the decision to develop a revised version of the ontology, and integrate it into OBI.

ONTIE 2.0 development required mapping the existing terms used in the IEDB to the BFO/OBI structure. Fig. 3 depicts a representative experiment to characterize a

T cell epitope. The terms captured in the IEDB to describe such experiments include the immunization procedure, ELISPOT assay, host, immunogen, antigen, antigen-presenting cells and effector cells. The first two of these correspond to different steps of the experiment, and are clearly processes as they occur over time. The others are objects playing specific roles. For example, the same type of protein can be the immunogen in the immunization protocol and the antigen in the ELISPOT assay. Similarly, the same type of cell type can play the role of an effector cell and antigen presenting cell. In other words, these are structurally the *same* objects playing different roles in different processes. Examples for distinct types of objects, on the other hand, are organisms, cells, or proteins. This is the primary breakdown of experimental terms in OBI: Experimental steps correspond to processes of which each is associated with certain roles or functions that can be played by certain types of objects.

This restructuring of the ontology was extended to the database schema. In the previous schema, each object-role combination was stored in a separate table, resulting in a great deal of redundancy. In the new schema, there is one table where all object-related information is stored and it can be referenced by other tables that associate objects with roles in a process. For example, a particular type of peptide can play the role of epitope, the role of immunogen and the role of antigen. Before restructuring, the same exact peptide would have been stored in the epitope table, the immunogen table and the antigen table, but is now only stored once in the object table and is referenced by its different roles. In addition to reducing redundancy, this makes it easier to extend the database when new object types are encountered.

Although the IEDB aims for a complete representation of the data, not all steps of an experiment are captured explicitly. In the T cell epitope identification experiment in Fig. 3, the IEDB will capture information about the immunization and ELISPOT assay protocols. Although other steps exist, these are not as relevant for an accurate representation of the epitope-related information derived from the experiment. Several steps are therefore only implicitly captured. For example, the use of mouse lymph node cells implies that the lymph nodes were removed and apportioned in some fashion. The information about those tissue harvesting and cell isolation protocols is not captured by the IEDB, but their occurrence is implied.

Relationships

In the initial IEDB ontology, the only connection between terms was the unspecific "has" relationship, e.g., immunization *has* a host. In ONTIE 2.0, the more specific BFO/OBI relations are used instead. For example, an organism *participates in* an immunization, and *plays role* host. Between objects, the most important relations are *has part* and *derived from*. For example, a T cell *has part* TCR or a cell extract is *derived from* cells.

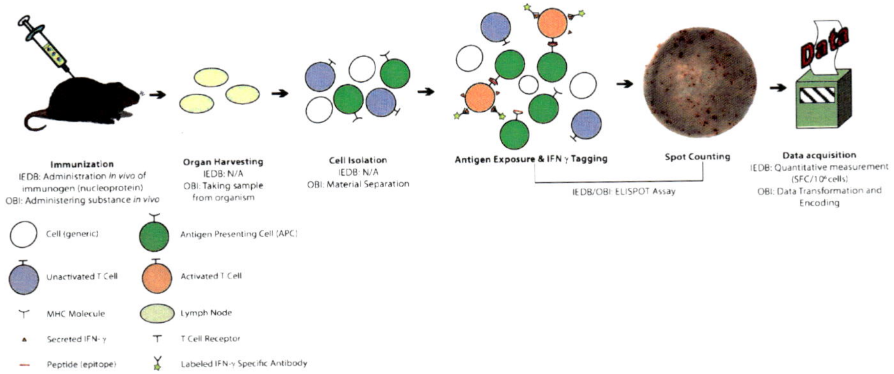

Fig. 3 A typical T cell epitope identification experiment captured by the IEDB. All protocols are described as in the IEDB. This experiment starts with an Immunization protocol with a naïve (in the immunological sense) mouse (playing the role of host) and SARS coronavirus nucleoprotein (playing the role of immunogen) as an input and an immunized mouse as an output. The immunized mouse is input for an organ harvesting protocol that produces mouse lymph nodes as output. Lymph cells are next generated as output from a cell isolation protocol. This yields a mixture of T cells and potential antigen presenting cells (APC) among other cell types. These protocols are labeled N/A for IEDB as they are not explicitly captured. In the next protocol (ELISPOT assay) a peptide that can be derived from the protein administered in the first protocol, is administered as the antigen, to the isolated cell mixture. APCs present bound peptide on their surface via MHC molecules. T cells specific for this peptide will bind to the peptide-MHC complex and become activated This results in the secretion of interferon γ (IFN-γ). Next, labeled antibodies specific for IFN-γ are added and bind to IFN-γ bound to the plate via a capture antibody (not depicted) in the vicinity of these activated T cells. In the next step – Spot counting, the plate is scanned by an instrument and spots are counted where labeled antibodies have bound. The final step – Data acquisition – is the transformation of the raw count to a frequency measurement followed by encoding into a digital format

In addition to the previously defined BFO relationships, we needed to introduce new ones to model all the terms encountered in the IEDB. A linear peptidic epitope is not necessarily *part of* or *derived from* a source protein molecule (Fig. 4). In many cases, the peptides used in an experiment are artificially synthesized and were therefore never part of any protein molecule. In these cases, the peptides have a special relationship termed "*proxy for*." A *proxy for* relationship indicates that the material is used in an experiment, observations are made, and conclusions are drawn on the material or class of materials that it is believed to represent. For example, the peptide in Fig. 4 is a *proxy for* a region of the SARS coronavirus nucleoprotein in the experiments in Fig. 3. In this experiment, observations are made on the peptide with the assumption that these observations can be generalized to any peptide with the same sequence.

A similar *proxy for* relationship is often encountered with experimental observations. In the ELISPOT assay in Fig. 3, the number of spots on a plate is counted and it is assumed that each of these represents one cell producing interferon-γ (IFN-γ). Thus, this count of spots is *proxy for* the number of spot forming cells (SFC).

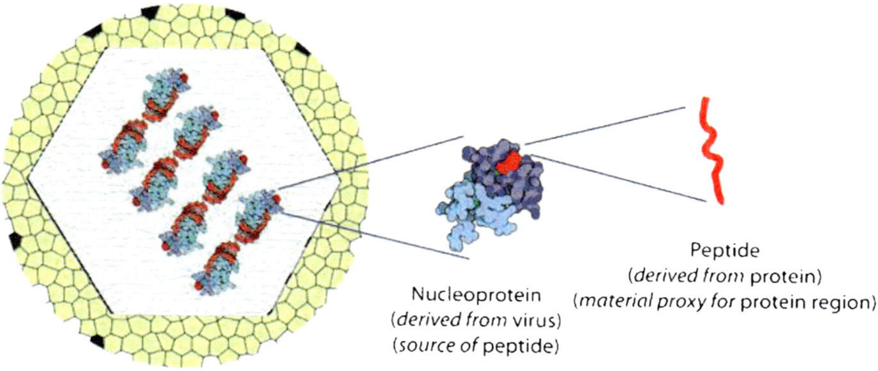

Fig. 4 *Objects and their relationships*. This figure illustrates the relationship among several objects. The SARS Coronavirus typically has many copies of the nucleoprotein in the virion. The nucleoprotein is derived from the virus and, conversely, the virus is the source of the nucleoprotein. The peptide is identical to a particular stretch of amino acids in the nucleoprotein and can therefore serve as material proxy for that region of the nucleoprotein in experiments. If it was created by digesting the nucleoprotein, then it would have a derived from/source of inverse relationship

These *proxy for* relations form a chain that can be traversed all the way back to the immune response. That is, the spot count is *proxy for* the SFC, which is *proxy for* the activation of T cells, which is *proxy for* the frequency of epitope specific T cells induced by the immunization protocol. The intermediate steps can be subsumed so that the count of spots is ultimately *proxy for* the frequency of epitope-specific T cells induced by the immunization protocol.

Summary and Conclusions

A vast amount of epitope information produced by experimental researchers is available in the scientific literature. Aside from the physical separation of this information in different publications, these experiments employ a wide variety of techniques, obtain different types of measurements, and are reported in divergent formats. Capturing this data in a consistent fashion enhances its utility and is what the IEDB team strives towards. This requires capturing the information in a natural, coherent, and unambiguous data structure. Developing this structure requires expertise from, and communication between, domain-expert curators and data modelers.

By applying the knowledge gained from the curation of experimental contexts that did not easily fit into our previous schema and utilizing concepts drawn from OBO, OBI, and BFO, we developed an ontology of immune epitopes for general use. Its development in accordance with OBI ensures its compatibility with the plethora of other ontologies developed under the OBO foundry. Simultaneously, we redesigned the IEDB data structure to follow the principles of the ontology and house a representation of the data that will be easier to maintain and extend. The finalized new version of the Immune Epitope Database based on this work, IEDB 2.0, is accessible to the public since 2009.

References

Ashburner M, Ball CA et al (2000) Gene ontology: Tool for the unification of biology. The Gene Ontology Consortium. Nat Genet 25(1):25–29

Grenon P, Smith B et al (2004) Biodynamic ontology: Applying BFO in the biomedical domain. Stud Health Technol Inform 102:20–38

Peters B, Sidney J et al (2005) The immune epitope database and analysis resource: From vision to blueprint. PLoS Biol 3(3):e91

Rosse C, Mejino JL Jr (2003) A reference ontology for biomedical informatics: The Foundational Model of Anatomy. J Biomed Inform 36(6):478–500

Salimi N, Vita R (2006) The biocurator: Connecting and enhancing scientific data. PLoS Comput Biol 2(10):e125

Sathiamurthy M, Peters B et al (2005) An ontology for immune epitopes: Application to the design of a broad scope database of immune reactivities. Immunome Res 1(1):2

Smith B, Ashburner M et al (2007) The OBO Foundry: Coordinated evolution of ontologies to support biomedical data integration. Nat Biotechnol 25(11):1251–1255

Vita R, Vaughan K et al (2006) Curation of complex, context-dependent immunological data. BMC Bioinform 7:341

Whetzel PL, Brinkman RR et al (2006a) Development of FuGO: An ontology for functional genomics investigations. Omics 10(2):199–204

Whetzel PL, Parkinson H et al (2006b) The MGED Ontology: A resource for semantics-based description of microarray experiments. Bioinformatics 22(7):866–873

TEPIDAS: A DAS Server for Integrating T-Cell Epitope Annotations

M. García-Boronat, C.M. Díez-Rivero, and Pedro Reche

Abbreviations

CMV	Cumulative phenotypic frequency
DAS	Distributed annotation system
HLA I	Human leukocyte antigen class I
PSSM	Position-specific scoring matrix

Introduction

Recent years have witnessed the birth of Immunoinformatics, an emerging subdiscipline of Bioinformatics. With the burgeoning explosion of immunological data, computational analysis has become an essential element of immunology research, facilitating the understanding of the immune function by modeling the interactions among immunological components (Petrovsky and Brusic 2006). Another major role in Immunoinformatics is the efficient management, storage, and annotation of such data. Following those principles, a large number of immunoinformatics resources, including immune-related databases and sophisticated analysis software, are available through the World Wide Web (Davies and Flower 2007). Collectively, these resources contribute to the advances made in immunological research. Yet, there is still a major step to be taken toward the integration of all these resources, as ideally, multiple research groups should be able to exchange and compare their data, in a quick and efficient fashion.

In this chapter, we show an example of how an epitope database can be integrated to other database resources using the Distributed Annotation System (DAS) (Dowell et al. 2001). For that we describe the TEPIDAS server, a DAS Annotation Server of HLA I-restricted CD8 T-cell epitopes specific of human pathogenic organisms.

P. Reche (✉)
Facultad de Medicina, Departamento de Immunología (Microbiología I), Universidad Complutense de Madrid, Pabellón 5°, planta 4ª, 28040, Madrid, Spain
e-mail: parecheg@med.ucm.es

D.R. Flower et al. (eds.), *Bioinformatics for Immunomics*, Immunomics Reviews 3,
DOI 10.1007/978-1-4419-0540-6_5, © Springer Science+Business Media, LLC 2010

The Distributed Annotation System

Introduction

The distributed annotation system defines a communication protocol used to exchange biological annotations from a number of heterogeneous distributed databases. The key idea behind the DAS concept is that annotations should not be provided by single centralized databases but instead be spread over multiple sites. This distribution of data encourages a divide-and-conquer approach to annotation, where experts provide and maintain their own annotations.

The Protocol

Currently, there are two versions of the DAS protocol. The original DAS protocol (DAS/1) was designed to serve annotation of genomic sequences. That protocol was later extended (DAS/2) to be applicable to alignments and 3D structure information (Prlic et al. 2005). It is very likely that further extensions of the protocol will appear in a near future, such as the new extension for electron microscopy data recently published by Macias et al. (2007).

The DAS protocol is a simple http-based client–server system. DAS clients make requests in the form of a URL to the servers and receive simple XML responses (Crook and Howell 2007). The architecture of the system will be next described in the following subsection.

The Architecture of the System

The basic system is composed of a reference server, one or more annotation servers, and an annotation viewer. The reference server is responsible for serving genome maps, sequences and information related to the sequencing process. Annotation servers are responsible for returning the annotations on a defined region (given a start and stop position coordinates) of the genome. The annotation viewer can either be a simple web browser, which will visualize the raw XML data provided by the server, or a graphical client such as the Center for Biological Sequence Analysis (CBS) DAS viewer (Olason 2005) accessible at http://www.cbs.dtu.dk/cgi-gin/das. This viewer translates the XML annotations to aligned graphical tracks making it easier to visualize the features along the length of the protein. Additional information about the annotations is shown in a pop-up window when the mouse points to an annotation track.

Although the servers are conceptually divided between reference and annotation servers, there is in fact no key difference between them. A single server can provide both reference sequence information and annotation information. The only functional difference is that the reference sequence server is required to serve the coordinate map and the raw DNA, while annotation servers have no such requirement. Our TEPIDAS server falls into the category of annotation servers.

The DAS Registry

The DAS Registry is a public server (http://www.dasregistry.org) dedicated to the registration, validation, and listing of worldwide DAS servers. One can browse the list of available DAS sources at the Registry, as well as register his own DAS server for public use. The Registry automatically validates the DAS server when it is being registered, ensuring that it returns well-formed XML responses. In addition, it periodically tests DAS sources and notifies their administrators if they are unavailable.

When you register your DAS server, you have to specify the Coordinate System of your source in order to describe the kind of data that are being made available. This information is important for the DAS clients to deal with data correctly, as they often can accept data served in multiple coordinate systems. The Coordinate System is described by the following four fields: "Authority," "(assembly) Version," "Type," and "Organism." The assembly version is important for genome assemblies, but not really applicable for other datasets like UniProt sequences; therefore, this field is optional. The "authority" is the name of an authority/institution that defines the accession codes of a coordinate system or that provides a gene build. In the latter case this field also contains the "version" number of the assembly. The "type" or category of the coordinate system refers to the physical dimension of the annotated data. Some examples include: Chromosome, Clone, Protein Sequence, and Protein Structure. The last field is the "organism" the data refer to. Not every DAS source is organism specific, and therefore this field is optional.

During the registration process, you also have to specify the capabilities of your DAS source, that is the types of queries that your server will be able to serve a response to. Some basic queries that can be used by a client to interrogate a DAS server are: "dna," "features," and "types." The "dna" query can be used to fetch a segment of DNA from a reference server. "features" is the query used to retrieve the actual annotations, and the "types" query returns a summary of the available annotation types. These three are just some examples of DAS queries. Readers can access the full list and specification of query types at the DAS web page (http://www.biodas.org).

The TEPIDAS server has been registered at the DAS registry since February 2008 and has the unique id DS_545. The coordinate system defined for TEPIDAS is Uniprot (Wu et al. 2006), as the "authority," and Protein Sequence, as the "type." As for TEPIDAS capabilities, our server implements the "types" and "features" queries. Note that our server is just an annotation server, and therefore it does not provide the "dna" query, served only by reference servers. A comprehensive description of the TEPIDAS server follows next.

TEPIDAS

TEPIDAS is a DAS annotation server that provides annotations for CD8 T-cell epitopes consisting of the distinct HLA I molecules to which that epitope binds, following the UniProt coordinates system. TEPIDAS is implemented using ProServer (Finn et al. 2007), a lightweight Perl-based DAS server that does not depend on a separate HTTP server. The annotations are precalculated and the results stored in a relational database, allowing for fast retrieval and update of data. When a client makes a query to the TEPIDAS server, ProServer simply retrieves the relevant information from the relational database and composes the XML response.

Annotations Served by TEPIDAS

TEPIDAS annotates CD8 T-cell epitopes according to the HLA I molecules that restrict them. Epitopes were obtained from the EPIMHC (Reche et al. 2005) and IMMUNEEPITOPE (Peters et al. 2005) databases, and were selected to be experimentally defined in humans infected with the pathogen or immunized with the relevant source antigen. HLA I-restriction annotations can be classified as experimental, when determined experimentally, or predicted. Predictions of the epitopes binding HLA I molecules were obtained using a set of 72 position-specific scoring matrices (PSSMs), also known as weight matrices of profiles, which are obtained from aligned peptides known to bind to the relevant HLA I molecules. This predictive method is described in full detail at (Reche et al. 2002, 2004). In addition to the experimental and predicted data, the cumulative phenotypic frequency (CMV) of the T-cell epitope HLA I restriction is also provided for five ethnic groups (Black, Caucasian, Hispanic, North American natives, and Asian). CMV was computed using the gene and haplotype frequencies of the relevant HLA I alleles (Reche et al. 2006). The potential population protection coverage of a T cell epitope-based vaccine is determined by the percentage of the population that could elicit a T cell response to the epitopes, which in turn is given by the CMV of HLA I molecules restricting these epitopes.

TEPIDAS Query Capabilities

As we mentioned before, TEPIDAS capabilities include the "types" and "features" queries. An explanation and an example for each query follow next.

The "types" query returns a list of all the distinct HLA I molecules that are used to annotate the epitopes. A total of 125 different HLA I restriction elements are included in TEPIDAS. To make this query to the server, you simply have to access the following URL through your web browser:

http://imed.med.ucm.es:9000/das/tepidas/types
and the XML response you will get is shown as follows.


```
<?xml version="1.0" standalone="no"?>
<!DOCTYPE DASTYPES SYSTEM "http://www.biodas.org/dtd/dastypes.dtd">
<DASTYPES>
  <GFF version="1.0" href="http://imed.med.ucm.es:9000/das/tepidas/types">
  &lt;SEGMENT version="1.0">
    <TYPE id="HLA-A*02" method="Experimental" category="default"></TYPE>
    <TYPE id="HLA-A*0201" method="Experimental" category="default"></TYPE>
                                  .
                                  .
                                  .
    <TYPE id="HLA-B*02706" method="Predicted" category="default"></TYPE>
    <TYPE id="HLA-B*02709" method="Predicted" category="default"></TYPE>
    <TYPE id="HLA-B*027" method="Predicted" category="default"></TYPE>
  &lt;/SEGMENT&gt;
</GFF>
</DASTYPES>
```


Only a part of the XML response file is shown due to length constraints. Each type has an "id" that corresponds to the name of the HLA I molecule. There is also a "method" attribute that distinguishes between experimental and predicted annotations. In addition, a third attribute named "category" can be used to group different types, although we have not used that attribute, and therefore *default* is the "category" shown in the response.

The other type of query supported by TEPIDAS is the "features" query, which returns the actual annotations made on a reference UniProt sequence. An annotation feature includes the following information: the start and end position of the feature annotated, the method used to annotate it (experimental or predicted), the type of the annotation (the HLA I molecule to which it binds), a link to the UniProt page of the reference protein sequence, and a note field with additional complementary information. The information on the note varies depending on the feature's method. Common fields in the note of both methods are: the epitope source species name and taxonomy identifier, the name of the source protein, the cumulative phenotypic frequency (CMV) of the T-cell epitope HLA-I restriction for five ethnic groups (Black, Caucasian, Hispanic, North American natives, and Asian), and the immunogen type. Specific fields for the features with an experimental "method" are: T-cell epitope activity assays, the experimental HLA I restriction element, its binding level (low, moderate, high, or unknown), and the predicted HLA I restriction elements. As for the features with a predicted "method" the note also includes the predicted HLA I restriction element, as well as an extended prediction with additional HLA I restriction elements for that epitope.

The "features" query has several arguments that can be optionally used to restrict the results. For example, the following URL string:

http://imed.med.ucm.es:9000/das/tepidas/features?segment=P26664

will return all the features annotated on the UniProt protein sequence identified with the accession number P26664 (which will also be the features id).

If we want to restrict our query to the annotations on a particular region of the protein sequence, we could use:

http://imed.med.ucm.es:9000/das/tepidas/features?segment=Q9WMX2:885,893

which returns all the features for the protein sequence with accession number Q9WMX2 that lie within the region defined by the start and end positions 885 and 893. The XML response to this query is shown as follows.


```
<?xml version="1.0" standalone="yes"?>
<!DOCTYPE DASGFF SYSTEM "http://www.biodas.org/dtd/dasgff.dtd">
<DASGFF>
  <GFF version="1.01" href="http://imed.med.ucm.es:9000/das/tepidas/features">
    &lt;SEGMENT id="Q9WMX2" version="1.0" start="885" stop="893">
    <FEATURE id="Q9WMX2" label="Q9WMX2">
      <TYPE id="HLA-A*2402" reference="no" subparts="no" superparts="no">
       HLA-A*2402</TYPE>
      <METHOD id="Experimental">Experimental</METHOD>
      <START>885</START>
      <END>893</END>
      <ORIENTATION>0</ORIENTATION>
      <NOTE>
       Epitope Source Species: Hepatitis C virus; TaxID: 11103
       Epitope Source Protein: Genome polyprotein
       T cell Epitope Activity positive on: 51 Chromium Release,
       Cytokine bioassay
       MHCI Restriction Element: HLA-A*2402 (Experimental)
       MHCI Binding level: unknown
       Predicted MHCI Restriction: HLA-A*24,  HLA-A*2402
       Cummulative Phenotypic Frequency of MHCI(%):
       5.5(Black),12.8(Caucasian),22.9(Hispanic),
       40.3(North American Natives),34.3(Asian)
       Immunogen: Infection</NOTE>
     <LINK href="http://www.ebi.uniprot.org/unipro t-
srv/uniProtView.do?proteinAc=Q9WMX2">http://www.ebi.uniprot.org/uniprot-
srv/uniProtView.do?proteinAc=Q9WMX2</LINK>
    </FEATURE>
    <FEATURE id="Q9WMX2" label="Q9WMX2">
      <TYPE id="HLA-A*24" reference="no" subparts="no" superparts="no">
       HLA-A*24</TYPE>
      <METHOD id="Predicted">Predicted</METHOD>
      <START>885</START>
      <END>893</END>
      <ORIENTATION>0</ORIENTATION>
      <NOTE>
       Epitope Source Species: Hepatitis C virus; TaxID: 11103
       Epitope Source Protein: Genome polyprotein
       T cell Epitope Activity: predicted
       MHCI Restriction Element: HLA-A*24 (Predicted)
       MHCI Binding level: unknown
       Extended predicted MHCI Restriction: HLA-A*24, HLA-A*2402
       Cummulative Phenotypic Frequency of MHCI(%):
       5.5(Black),12.8(Caucasian),22.9(Hispanic),
       40.3(North American Natives),34.3(Asian)
       Immunogen: Infection</NOTE>
     <LINK href="http://www.ebi.uniprot.org/unipro t-
srv/uniProtView.do?proteinAc=Q9WMX2">http://www.ebi.uniprot.org/uniprot-
srv/uniProtView.do?proteinAc=Q9WMX2</LINK>
    </FEATURE>
    &lt;/SEGMENT&gt;
  </GFF>
   </DASGFF>
```


Example: Access TEPIDAS from the SPICE Graphical Client

In the previous section we have described how to access TEPIDAS annotations using formatted queries from a web browser, and we have also shown examples of the XML responses to the queries. We will now describe a different way of accessing TEPIDAS from a graphical client such as SPICE (Prlic et al. 2005). We hope that this example will illustrate the integration capability of DAS.

SPICE is a Java program that can be used to visualize annotations of protein sequences and protein structures. It is available at: http://www.efamily.org.uk/software/dasclients/spice. SPICE accepts either a PDB (Berman 2008) or a UniProt code, and integrates information from four different types of DAS servers: (1) a protein sequence server that provides the sequence (typically UniProt), (2) an alignment server that provides the alignment between the protein sequence and its structure, (3) a structure server that serves the 3D coordinates displayed, and (4) several feature servers that provide precalculated annotations, as for example TEPIDAS among others.

The SPICE viewer window consists of (1) a left structure panel, which provides a 3D visualization of the molecule using the open source Jmol library (http:/www.jmol.org), and (2) a right 2D feature panel that displays the annotations provided by the distributed servers. This is illustrated in Fig. 1 using the protein sequence with UniProt code P35961 as an example. As we can appreciate in Fig. 1, SPICE has automatically mapped that protein sequence to PBD "19GN" using its default alignment server. Figure 1 clearly shows how different annotations from several DAS servers can be integrated and collectively visualized through a graphical client such as SPICE. Users can choose which DAS annotations servers to use, as well as add new local DAS sources that are still under development or have not been registered with the DAS registry.

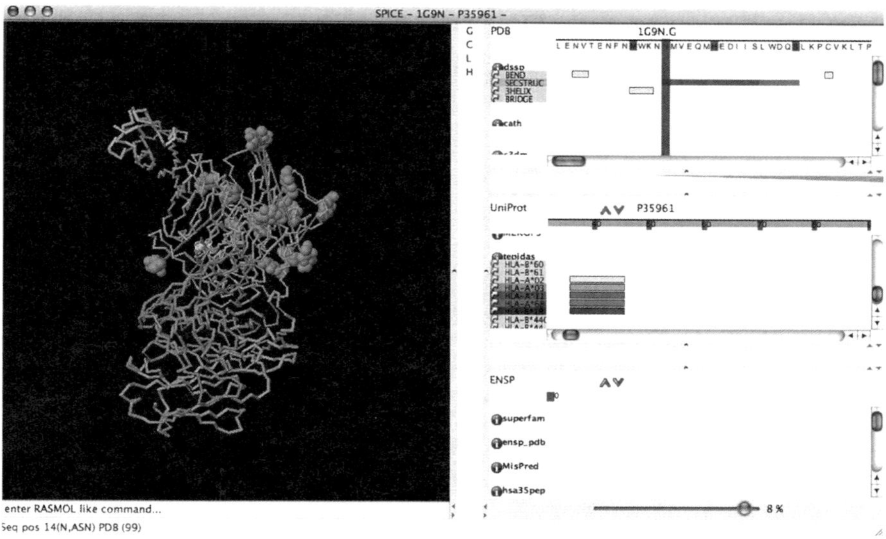

Fig. 1 SPICE viewer window. *Left panel* provides a 3D visualization of the molecule. *Right panel* displays the annotations provided by the distributed serves. This figure was generated using the UniProt code P35961 as the reference sequence. SPICE's alignment server automatically maps the protein sequence to a 3D structure (1G9N in this example). Feature annotations from TEPIDAS are displayed in the *right center panel* as *rectangular* tracks colored as the HLA I molecules on their *left* under the tepidas source descriptor

SPICE retrieves the protein sequence pertaining to the selected UniProt code and displays it as a ruler with relative position numbers, although there is a zoom feature that allows it to be expanded up to amino acid level as shown in Fig. 2 TEPIDAS annotation features are listed below the sequence in that figure. On the left of the panel, below the "tepidas" descriptor, appears the type of HLA I molecule of the corresponding feature shown as a colored rectangle on the right. When the user clicks on a feature, a pop-up window appears, containing all the information of the feature, including the explanatory note. In addition, the PDB coordinates of the selected feature will be highlighted at the left panel, enabling the location of the epitope at the 3D structure. Figure 3 shows an example of a pop-up window with feature information.

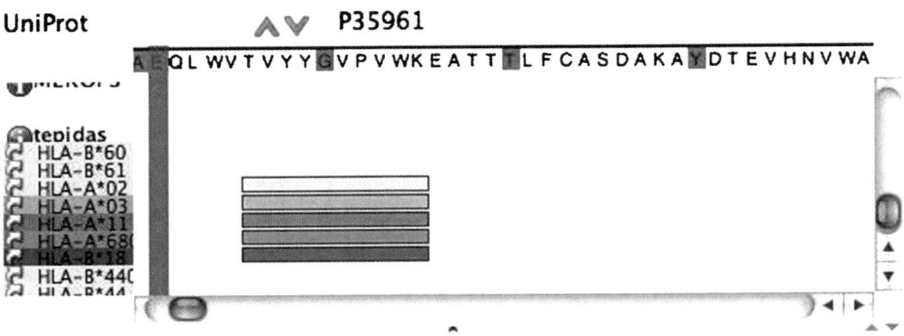

Fig. 2 SPICE zooming capability. Protein sequence visualized at amino acid level

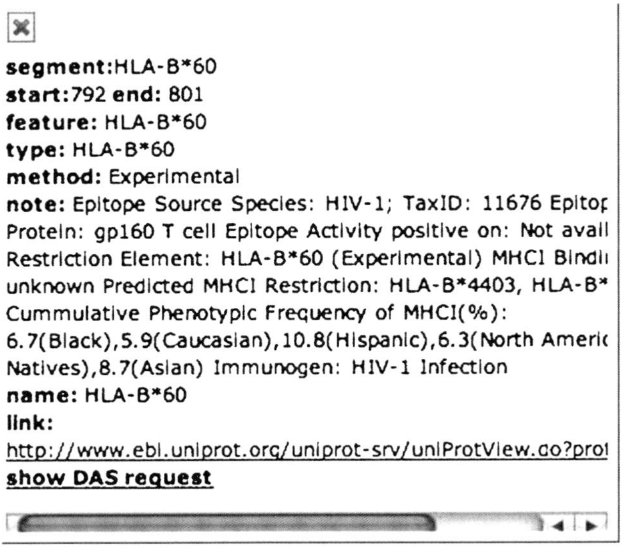

Fig. 3 Pop-up window containing all the information for feature HLA-B*60 annotated for protein sequence referenced by UniProt code P35961

Conclusion

DAS is an important, simple, and yet a powerful system for exchanging and viewing biological data that are already being used in real-world bioinformatics applications. The TEPIDAS annotation server described in this chapter is a clear example of how epitope data can be integrated and shared by the research community using the DAS architecture. The complexity of immune interactions and the data-intensive nature of immune research make Immunoinformatics a suitable area that could greatly benefit from the advantages of using such a powerful integration and annotation system, allowing to gain a more insightful understanding of the complexities of the immune system.

Acknowledgments We would like to thank Alfonso Valencia, Osvaldo Graña, and Jaime Fernandez Vera from the Spanish National Cancer Research Center (CNIO) for their helpful advice on DAS and ProServer. Work and authors were supported by grant SAF2006-07879 from the "Ministerio de Educación y Ciencia" of Spain, granted to PR.

References

Berman HM (2008) The Protein Data Bank: a historical perspective. Acta Crystallogr A 64(Pt 1): 88–95

Crook SM, Howell FW (2007) XML for data representation and model specification in neuroscience. Methods Mol Biol 401:53–66

Davies MN, Flower DR (2007) Harnessing bioinformatics to discover new vaccines. Drug Discov Today 12(9–10):389–395

Dowell RD, Jokerst RM et al (2001) The distributed annotation system. BMC Bioinformatics 2:7

Finn RD, Stalker JW, Jackson DK et al (2007) ProServer: a simple, extensible Perl DAS server. Bioinformatics 23(12):1568–1570

Macias JR, Jimenez-Lozano N, Carazo JM (2007) Integrating electron microscopy information into existing Distributed Annotation Systems. J Struct Biol 158(2):205–213

Olason PI (2005) Integrating protein annotation resources through the Distributed Annotation System. Nucleic Acids Res 33(Web Server issue):W468–W470

Peters B, Sidney J et al (2005) The immune epitope database and analysis resource: from vision to blueprint. PLoS Biol 3(3):e91

Petrovsky N, Brusic V (2006) Bioinformatics for study of autoimmunity. Autoimmunity 39(8):635–643

Prlic A, Down TA, Hubbard TJ (2005) Adding some SPICE to DAS. Bioinformatics 21(Suppl 2): ii40–ii41

Reche PA, Glutting JP, Reinherz EL (2002) Prediction of MHC class I binding peptides using profile motifs. Hum Immunol 63(9):701–709

Reche PA, Glutting JP et al (2004) Enhancement to the RANPEP resource for the prediction of peptide binding to MHC molecules using profiles. Immunogenetics 56(6):405–419

Reche PA, Zhang H et al (2005) EPIMHC: a curated database of MHC-binding peptides for customized computational vaccinology. Bioinformatics 21(9):2140–2141

Reche PA, Keskin DB, Hussey RE et al (2006) Elicitation from virus-naïve individuals of cytotoxic T lymphocytes directed against conserved HIV-1 epitopes. Med Immunol 5:1

Wu CH, Apweiler R, Bairoch A et al (2006) The Universal Protein Resource (UniProt): an expanding universe of protein information. Nucleic Acids Res 34(Database issue):D187–D191

Databases and Web-Based Tools for Innate Immunity

Sneh Lata and G.P.S. Raghava

Introduction

Immune system has two arms: The faster acting innate immune system and the slower developing adaptive immune system. The innate immune system though being the universal and ancient form of host defense was thought to merely provide a necessary first line of defense as it was considered to be nonspecific. In contrast, adaptive immunity, considered to be the main player in host defense, is highly specific and generates long-lasting immunological memory (Gourley et al. 2004). This "immunological memory," is the foundation upon which rests the concept of vaccination. Therefore, most of the research activity, till date, was focused on understanding the adaptive immunity and discovering vaccine candidates that could activate the adaptive immune system and generating a memory for the pathogenic encounter.

In the initial years, the vaccines used were either heat-killed or attenuated whole pathogen but over a period of time, the strategies to develop vaccines changed tremendously i.e., from whole pathogen to antigens and from antigens to antigenic regions or epitopes. These antigenic regions or subunit vaccines have numerous advantage over vaccines based on whole organism. Therefore, research in the field of subunit vaccine designing has been directed mainly to fish out new subunit vaccine candidate from pathogens. Identification of potential vaccine candidates continued to be a major challenge for experimentalists, in subunit vaccine designing. For this reason, a large number of bioinformatics databases (Saha et al. 2005; Huang and Honda 2006; McSparron et al. 2003; Bhasin et al. 2003; Peters et al. 2005) and tools (Pellequer and Westhof 1993; Rammensee et al. 1997; Singh and Raghava 2001; Reche et al. 2002; Bhasin and Raghava 2004a, b; Saha and Raghava 2004; Bhasin and Raghava 2005; Saha and Raghava 2006) have been developed for predicting antigenic epitopes or antigens. However, only limited success has been achieved so far, as epitopes/antigens show poor immunogenicity, thereby defeating the main goal of vaccination i.e., to provide strong and long-lasting immunity.

G.P.S. Raghava (✉)
Institute of Microbial Technology, Sector 39A, Chandigarh, India
e-mail: raghava@imtech.res.in

D.R. Flower et al. (eds.), *Bioinformatics for Immunomics*, Immunomics Reviews 3,
DOI 10.1007/978-1-4419-0540-6_6,

This clearly indicates that subunit vaccines designed based on adaptive-immunity is not sufficient to elicit a robust and sustained immune response.

Accumulating evidences now suggest that addition of components of innate immunity along with these epitopes helps to generate a strong and effective immune response and play a fundamental role in influencing immunological memory. The innate immune system acts as an activator and controller of the adaptive immune system (Iwasaki and Medzhitov 2004). It provides critical signals for the development of specific adaptive immune response (Janeway 1989; Fearon and Locksley 1996), which provide crucial information about the origin of antigen and direct the adaptive immune system as to what type of response shall be generated to combat it. Innate immune system employs a wide variety of receptors called pattern-recognition receptors (PRRs, which are relatively invariable) that detect evolutionarily conserved molecular patterns from pathogens (Janeway 1989; Barton and Medzhitov 2002). These patterns are termed pathogen- associated molecular patterns (PAMPs). PAMPs are unique to the microbes and are rarely produced in the host system. PAMPs are not subject to mutation either as these are essential for the physiology and pathogenesis of the microbes. Thus it is clear now that innate immune system is endowed with the ability to discriminate between self and nonself to some extent and is not nonspecific. However, PRRs recognize endogenous ligands in host as well in conditions of stress. These receptors, therefore, play a dual role in host defense and homeostasis (Gordon 2002).

Moreover, engagement of PRRs by their ligands leads to activation of adaptive antigen-recognition receptors and induction of key co-stimulatory molecules, antimicrobial peptides and cytokines (Fearon and Locksley 1996) as well as maturation and migration of other cells. Overall, these cascade of events leads to the development of a robust and durable adaptive immune response (Schellack et al. 2006). Thus, the PAMPs can be used as adjuvants to be administered along with subunit vaccines in order to enhance its efficacy. Therefore, understanding the molecular mechanisms responsible for PAMPs and generation of downstream signaling effecter molecules would be crucial for the development of new approaches to vaccine formulation and immunotherapy. Although in the last decade progress has been made by the scientists to understand the innate immune system, bioinformatics-based research for innate immunity is still in its infancy. Some databases have been created that hold the detailed information about the innate immunity components and methods have been developed that can predict important proteins and peptides of innate immune system. In this chapter, we describe the databases and tools developed for innate immunity.

Databases for Innate Immune System

The Innate Immune Database

This is a database of innate immunity that contains computationally predicted transcription factor binding sites and related genomic features for a set of over 2,000 murine immune genes of interest (Korb et al. 2008). This database also includes

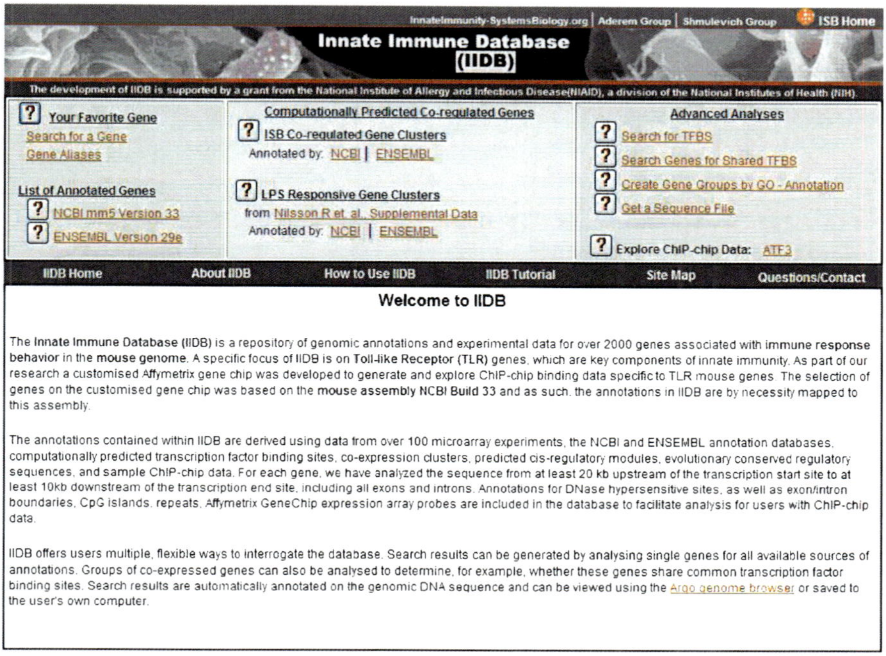

Fig. 1 A snapshot of IIDB homepage

microarray co-expression clusters and a host of web-based query, analysis and visualization facilities. This database provides the means to analyze individual genes or an entire genomic locus. It provides a broad resource to the research community, and a stepping stone towards the delineation of the network of transcriptional regulatory interactions underlying the integrated response of macrophages to pathogens. This database (Fig. 1) can be accessed free on the internet through http://db.systemsbiology.net/IIDB.

Innate Immunity Interactions Database

Innate DB is a publicly available database (Fig. 2) of genes, proteins, interactions, and pathways involved in innate immunity. It integrates known interactions and pathways from public databases and manually curated data into a centralized resource, which include data on >100,000 human and mouse interactions, cross-references to innate immunity relevant pathways, and detailed annotations from a variety of sources. This database also provides several bioinformatic tools to facilitate systems-level investigations of the innate immune response. These include the ability to upload expression datasets, which can be integrated with network/pathway data to investigate changes in gene expression in a network of interest.

Fig. 2 A snapshot of Innate DB homepage

The database also provides orthology predictions for human, mouse and bovine genes to facilitate the construction of orthologous networks in different species. This database is available freely at http://innatedb.ca/.

Pattern Recognition Receptor Database

The Pattern Recognition Receptor Database (PRRDB) is a comprehensive database that summarizes the detailed information about the pattern-recognition receptors and their ligands (Lata and Raghava 2008a, b). This database is first of its kind as information about PRRs and their corresponding binding PAMPs have never been compiled at one place before. The current version of the PRRDB (Fig. 3) has information of 500 pattern recognition receptors belonging to seven different classes. These include the detailed information about Toll-like receptors, Scavenger receptors, Mannose receptors, C-type lectin-like domain family members, DC-SIGN, PGRPs and Nucleotide-binding oligomerization domain (NOD) proteins (e.g., Fig. 4). These receptors belong to about 77 distinct organisms (vertebrates) ranging from insects to homo sapiens. Information about the PRRs includes their common names, synonymous names, organisms they belong to, sequence length and protein sequence. These receptors are further classified as single-pass type I/II membrane proteins, multi-pass membrane proteins, cytoplasmic or secreted proteins depending

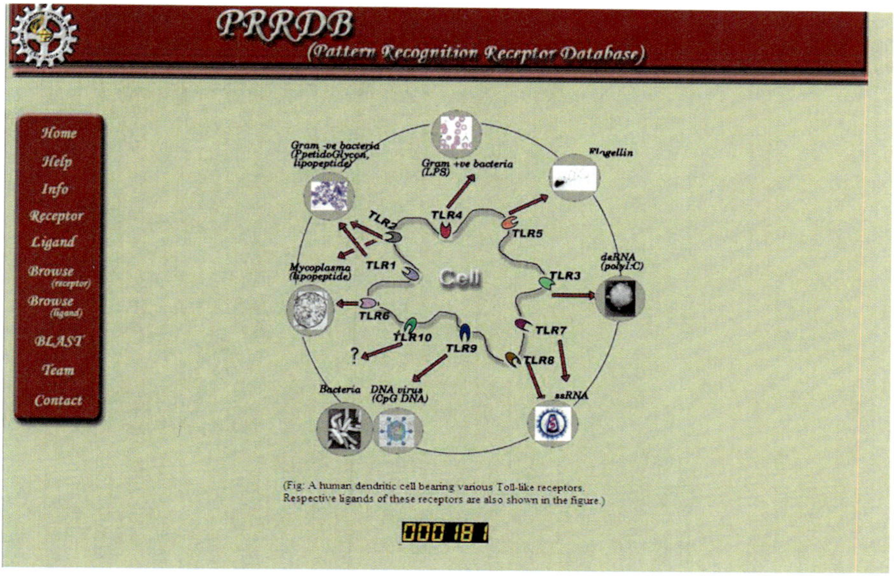

Fig. 3 A snapshot of Innate DB homepage

PRR ID	100001
Swiss-Prot ID	Q9BXR5 please click here to get the sequence in fasta format
Common name	TLR 10
Synonymous name	Toll-like receptor 10 precursor;CD290 antigen
Organism	Homo sapiens;Human
Sequence length	811
Type of receptor	single pass Type I membrane protein

Fig. 4 Detailed information for TLR10

upon their site of expression. Each entry also bears a Swiss-Prot ID that is hyperlinked to its corresponding entry in Swiss-Prot. The USP of the database is that it also contains information about 228 PAMPs. The information about PAMPs includes the name of the ligand, its source, receptor it binds to and whether the ligand is exogenous or endogenous (Fig. 5). As these ligands belong to a wide range of molecules, for convenience they were classified into different biochemical groups i.e., nucleic acid, protein, carbohydrates, glycoproteins, lipopeptides etc. From the PRRDB the users can utilize a number of online web tools that allow users to retrieve and analyze data. The database provides, along with other important details, the chemical structures of the ligands (or textual description in case structure is not available) (Fig. 6). This would help the scientists to look into the fine structural details that may be required for a PAMP to act as an adjuvant. One of the main problems for using PAMPs as adjuvants is the toxic effect exhibited by most of them. Therefore, the data collected include the information about a parent PAMP that shows toxic effect

Fig. 5 A snapshot of Innate DB homepage

Receptor Name	**TLR7**
Ligand	CL075
Source of ligand	**thiazoloquinolone derivative**
Reference	gorden05 levy06 gorski06

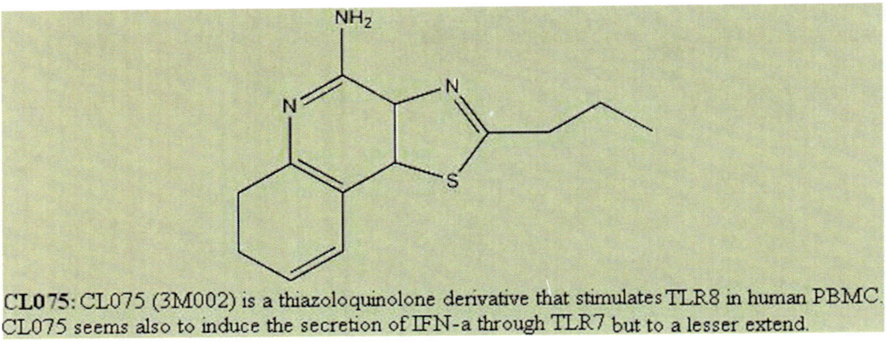

Fig. 6 A snapshot of Innate DB homepage

when used as adjuvant as well as its modified derivatives that is much safer to use. For example, it's known that LPS binds to TLR4 and can act as an adjuvant but it is toxic and pyrogenic whereas, its modified derivative MPL is nonpyrogenic and retains the TLR4- activating property. Examples like these would help the scientists to look into the structural details and study what is the minimal region required for the desired activity and is safer at the same time. Given the importance of these ligands as immune potentiators or adjuvant, PRRDB can prove to be a valuable resource for immunologists to understand immune system and in sub-unit vaccine designing. This database is available to the public at http://www.imtech.res.in/raghava/prrdb or http://bioinformatics.uams.edu/raghava/prrdb.

Tools for Innate Immunity

CTKPred

Cytokines are messenger molecules of immune system. They are small secreted proteins that mediate and regulate the immune system, inflammation and hematopoiesis. Recent studies have revealed important roles played by the cytokines in adjuvants as therapeutic targets and in cancer therapy. CTKPred is a web server for the prediction of family and subfamily of the cytokine superfamily proteins based on support vector

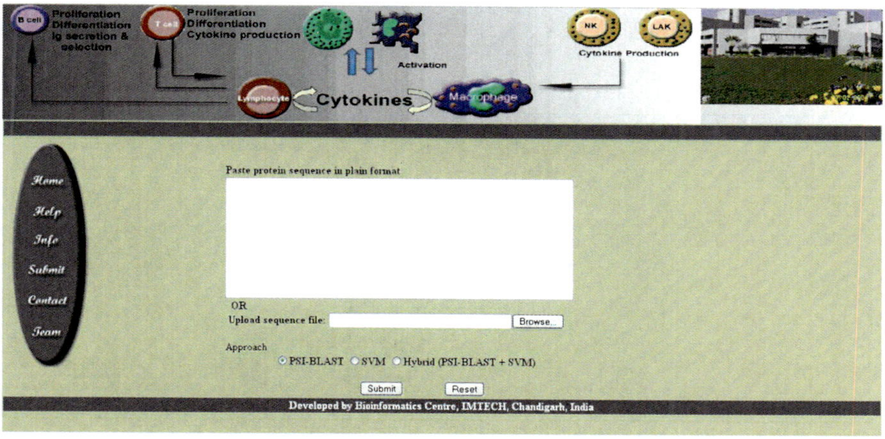

Fig. 7 CytoPred submission form

machine (Huang et al. 2005). The method classified cytokines and noncytokines with an accuracy of 92.5% by sevenfold cross-validation. The method is further able to predict seven major classes of cytokine with an overall accuracy of 94.7%. A server for recognition and classification of cytokines based on multi-class SVMs has been set up at http://bioinfo.tsinghua.edu.cn/_huangni/CTKPred/.

CytoPred

In CytoPred (Fig. 7) a hybrid approach (PSI-BLAST+Support Vector Machine-based) is adopted to develop the prediction methods (Lata and Raghava 2008a, b). CytoPred first predicts whether a given test protein is a Cytokine or not. This is the superfamily classification and CytoPred is capable of predicting cytokines superfamily with an accuracy of 98.29%. If the test or query protein is predicted to be a cytokine then CytoPred predicts the class to which this query protein belongs to. In case the query protein goes to TGF-B family then subfamily of the query protein is also predicted (Fig. 8). The overall accuracy of classification of cytokines into four families and further classification into seven subfamilies is 99.77% and 97.24%, respectively. It has been shown by comparison that CytoPred performs better than the already existing CTKPred. A user-friendly server CytoPred has been developed and is available at http://www.imtech.res.in/raghava/cytopred.

AntiBP

Antibacterial peptides are important components of the innate immune system, used by the host to protect itself from different types of pathogens. Over the last few

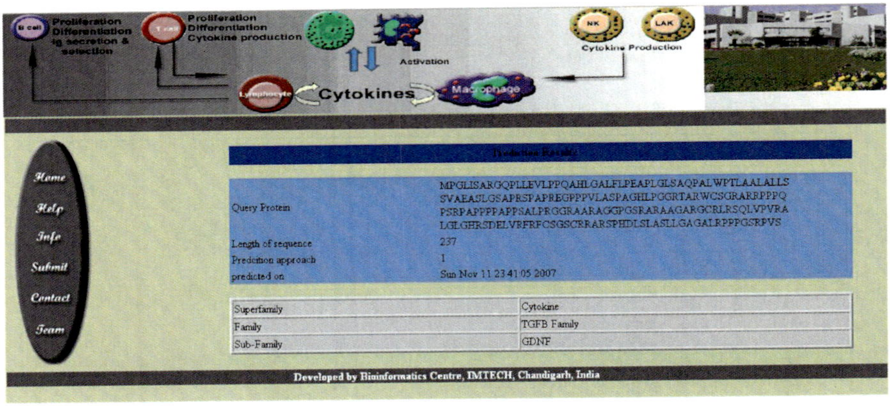

Fig. 8 CytoPred query result page

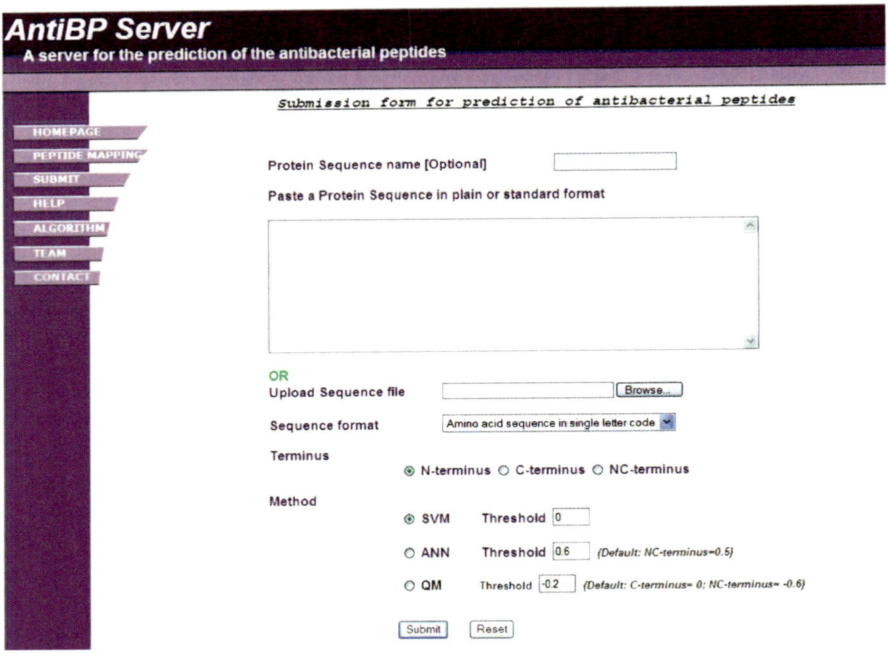

Fig. 9 A snapshot of AntiBP submission form

decades, the search for new drugs and drug targets and the growing number of antibiotic resistant bacteria has prompted an interest in these antibacterial peptides. The design of novel peptides with antimicrobial activities requires the development of methods for narrowing down the candidate peptides so as to enable rational experimentation by wet-lab scientists. AntiBP is a systematic attempt to understand this important class of peptides and to develop an algorithm for predicting antibacterial

Prediction result in tabular format

PEPTIDE	START POSITION	SCORE	Antibacterial Activiy
RLKSGKRKLMNSTRP	1	1.119	YES

Fig. 10 AntiBP query result page

peptides with high accuracy. AntiBP is a support vector machine (SVM)-, quantitative matrix (QM)- and artificial neural network (ANN)- based method meant to predict and thus discover efficacious antibacterial peptides (Lata et al. 2007). It was observed that particular types of residues are preferred over others at both the N and the C-termini of the antibacterial peptide. The models have been developed therefore, for the N-terminus, C-terminus and N+C terminus sequences. A user has just to provide a peptide sequence in single letter amino acid code as input. The user can select the approach to be used for prediction (Fig. 9). Users can also specify if the prediction is to be made from N-terminus or C-terminus. The prediction result is represented as shown in Fig. 10. The AntiBP server also has a provision to map a query peptide on to existing antibacterial peptides. The mapping option would help the scientists to find out if there is any stretch of antibacterial peptide existing in their long protein sequence. This method is likely to help the researchers in finding and in designing better peptides-based antibiotics. The user-friendly server is available to the public at http://www.imtech.res.in/raghava/antibp.

Conclusion

With the importance of the role played by innate immune system in body defense, the information in the databases developed for innate immune system would be very helpful for the wet lab immunologists to understand the innate immune system. The structures of ligands provided would help in the designing of novel and improved adjuvants and in turn, more efficient sub unit vaccines. The tools for the innate immune system would help in the prediction and classification of molecules that are important components of innate immune system. We hope, over a period of time, information on innate immunity will increase and so would grow the number of databases and methods for innate immune system.

References

Barton GM, Medzhitov R (2002) Toll-like receptors and their ligands. Curr Top Microbiol Immunol 270:81–92

Bhasin M, Raghava GPS (2004a) SVM based method for predicting HLA-DRB1 binding peptides in an antigen sequence. Bioinformatics 20:421–423

Bhasin M, Raghava GPS (2004b) Prediction of CTL epitopes using QM, SVM and ANN techniques. Vaccine 22:3195–3204

Bhasin M, Raghava GPS (2005) Pcleavage: An SVM-based method for prediction of constitutive proteasome and immunoproteasome cleavage sites in antigenic sequences. Nucleic Acids Res 33:W202–W207

Bhasin M, Singh H, Raghava GPS (2003) MHCBN: A comprehensive database of MHC binding and non-binding peptides. Bioinformatics 19:665–666

Fearon DT, Locksley RM (1996) The instructive role of innate immunity in the acquired immune response. Science 272:50

Gordon S (2002) Pattern recognition receptors: Doubling up for the innate immune response. Cell 111:927–930

Gourley TS et al (2004) Generation and maintenance of immunological memory. Semin Immunol 16:323–333

Huang J, Honda W (2006) CED: A conformational epitope database. BMC Immunol 7:7

Huang N, Chen H, Sun Z (2005) CTKPred: An SVM-based method for the prediction and classification of the cytokine superfamily. Protein Eng Des Sel 18(8):365–368

Iwasaki A, Medzhitov R (2004) Toll-like receptor control of the adaptive immune responses. Nat Immunol 5:987–995

Janeway CA Jr (1989) Approaching the asymptote? Evolution and revolution in immunology. Cold Spring Harbor Symp Quant Biol 54:1–13

Korb M, Rust AG, Thorsson V, Battail C, Li B, Hwang D, Kennedy KA, Roach JC, Rosenberger CM, Gilchrist M, Zak D, Johnson C, Marzolf B, Aderem A, Shmulevich I, Bolouri H (2008) The innate immune database (IIDB). BMC Immunol 9:7

Lata S, Raghava GP (2008a) PRRDB: A comprehensive database of pattern-recognition receptors and their ligands. BMC Genomic 9:180

Lata S, Raghava GP (2008b) CytoPred: A server for prediction and classification of cytokines. Protein Eng Des Sel 21(4):279–282

Lata S, Sharma BK, Raghava GPS (2007) Analysis and prediction of antibacterial peptides. BMC Bioinform 8:263

McSparron H et al (2003) JenPep: A novel computational information resource for immunobiology and vaccinology. J Chem Inf Comput Sci 43:1276–1287

Pellequer JL, Westhof E (1993) PREDITOP: A program for antigenicity prediction. J Mol Graph 11:204–210

Peters B et al (2005) The immune epitope database and analysis resource: From vision to blueprint. PLoS Biol 3:e91

Rammensee HG, Bachman J, Stevanovich S (1997) MHC ligands and peptide motifs. Landes Bioscience, Georgetown, pp 1–462

Reche PA, Glutting J, Reinherz EL (2002) Prediction of MHC class I binding peptides using profile motifs. Hum Immunol 63:701–709

Saha S, Raghava GPS (2004) BcePred: Prediction of Continuous B-Cell Epitopes in Antigenic Sequences Using Physico-chemical Properties. In: Nicosia G, Cutello V, Bentley PJ,Timis J (eds) ICARIS, LNCS 3239. Springer, Heidelberg, pp 197–204

Saha S, Raghava GPS (2006) Prediction of continuous B-cell epitopes in an antigen using recurrent neural network. Proteins 65(1):40–48

Saha S, Bhasin M, Raghava GPS (2005) Bcipep: A database of B-cell epitopes. BMC Genomics 6:79

Schellack C et al (2006) IC31, a novel adjuvant signaling via TLR9, induces potent cellular and humoral immune responses. Vaccine 24: 5461–5472

Singh H, Raghava GPS (2001) ProPred: Prediction of HLA-DR binding sites. Bioinformatics 17:1236–1237

Structural Immunoinformatics: Understanding MHC-Peptide-TR Binding

Javed Mohammed Khan, Joo Chuan Tong, and Shoba Ranganathan

Introduction

In higher jawed vertebrates, antigen presentation and recognition occurs in two steps, where, the peptide ligand first binds to the major histocompatibility complex (MHC) molecule followed by recognition of this peptide–MHC (pMHC) complex by the T cell receptor (TR). These two steps play a key role in the activation of normal adaptive immune responses. The first step in TR-mediated immune response is thus the binding and presentation of antigenic endogenous or exogenous peptide epitopes, which can be successfully predicted using sequence-based methods for alleles with large datasets of known binding peptides (Tong et al. 2007a). The second step, however, is an intricate theoretical problem that remains unsolved and is the next frontier in Immunoinformatics.

With the development of new structural modeling and docking techniques and an increase in the number of protein structures deposited in the Protein Data Bank (PDB) (Berman et al. 2000), the use of structure-based approaches to predict potential T-cell epitopes is increasingly successful (Ranganathan et al. 2008), often producing modeled structures accurate to within 2.00 Å RMSD from the experimental crystal structure, providing a wealth of information for structural analysis and prediction. With the development of a fast and accurate docking protocol followed by quantitative predictions for both MHC Class I and Class II alleles even with limited binding peptide data, we have been successful in unraveling the mystery behind the first step (Tong et al. 2004) in adaptive immune response. For an MHC molecule to recognize antigenic peptides and for pMHC to be subsequently recognized by TR, geometric and electrostatic complementarity between the receptor and ligand are essential, determining the stability of the complex. In this context, the introduction of structural information can greatly facilitate our understanding

S. Ranganathan (✉)
Department of Chemistry and Biomolecular Sciences and ARC Centre of Excellence in Bioinformatics, Macquarie University, 2109 Sydney, NSW Australia
Department of Biochemistry, Yong Loo Lin School of Medicine, National University of Singapore, Medical Drive, Singapore 117597

D.R. Flower et al. (eds.), *Bioinformatics for Immunomics*, Immunomics Reviews 3, DOI 10.1007/978-1-4419-0540-6_7,

of how well a pMHC complex can associate with TR besides being able to predict epitopes, capable of eliciting TR response.

Here we introduce structural immunoinformatics concepts of TR/pMHC interaction based on their three-dimensional experimental and modeled structures, toward the development of a predictive model. To start with, we briefly describe the structural characteristics of pMHC complexes (Tong et al. 2006a), from which a novel supertype classification for class I MHC alleles has been developed (Tong et al. 2007b). Based on the analysis of these characteristics, we have developed a rapid and precise docking protocol to generate models of pMHC complexes which has been applied to antigenic epitope prediction for specific alleles implicated in different diseases (Tong et al. 2006b, c; Tong et al. 2007c), where sequence-based approaches are inapplicable due to limited data on antigenic peptides. We then summarize the available TR/pMHC structural resources from the Internet. Finally, we extend our pMHC interaction parameters and highlight the importance of TR/pMHC interactions to decipher how antigenic peptides elicit a T cell response.

MPID-T and Structurally Derived Interaction Parameters

With the growth in the numbers of pMHC and TR/pMHC structures in PDB and some interaction parameters being reported as significant for peptide/MHC interactions (Kangueane et al. 2001), there was an increasing need for a database dedicated to these structures and their analysis. Hence a preliminary database called MPID (Govindarajan et al. 2003) was developed which is composed of 86 classical pMHC structures. This was later extended to the TR level when all the TR/pMHC structures along with the new pMHC structures were included. The new version was called MPID-T (Tong et al. 2006a) and it contained 187 pMHC and 16 TR/pMHC structures. MHC–peptide Interaction Database – TR (MPID-T) – is a manually curated database comprising all the X-ray crystallographic structures of pMHC and TR/pMHC complexes obtained from the PDB. It was developed in order to understand the structural determinants of TR/pMHC recognition and binding. It provides sequence-structure-function information governing TR/pMHC interactions besides a web-interface to carry out structural analysis of these complexes. The database (April 2008 update) contains over 294 pMHC complexes (Class I complexes: 235, Class II complexes: 59, TR/pMHC complexes: 42) from five MHC sources (human 187, murine 101, rat 3, chicken 2, and monkey 1), spanning 52 alleles. The analysis of these complexes revealed the significance of some protein–protein interaction parameters for the characterization of pMHC interface. These parameters include interface area or change in solvent accessible surface area (hydrophobic contacts), intermolecular hydrogen bonds, gap index (electrostatic interactions), and gap volume (geometric complementarity) between the MHC receptor and its corresponding peptide ligand. We will explain each of these parameters briefly with the view to extending them to characterize TR/pMHC interactions and thus help us determine the potential of a pMHC complex to bind to TR and thereby elicit T cell response.

Interface Area Between Peptide and MHC

One of the most significant forces that drive protein folding and protein–protein interactions are hydrophobic interactions. The hydrophobic free energy of a protein when it is transferred from polar to hydrophobic environment and the change in solvent accessible surface area (ΔASA) upon complex formation share a linear correlation (Chothia and Janin 1975). Thus, an indication of the binding strength of the interacting partners is provided by the knowledge of the surface area of a complex interface in direct contact with solvent. The measure of the maximum permitted van der Waals' contact area that is covered by the center of a water molecule as it rolls over the surface of the protein gives the accessible surface area. Interface area for pMHC complexes is defined as the mean ΔASA on complex formation when going from separated MHC and peptide molecules to a pMHC complex state and is calculated as half the sum of the total ΔASA of both molecules for each type of complex (Tong et al. 2006a). The interface areas of all the molecular systems are computed using the program NACCESS (Hubbard and Thornton 1993).

$$\Delta \mathrm{ASA(pMHC)} = [\mathrm{ASA(MHC)} + \mathrm{ASA(peptide)} - \mathrm{ASA(pMHC)}]/2 \qquad (1)$$

The mean ΔASA for class I pMHC complexes was found to be $903.30 \pm 260.90\,\text{Å}^2$. Similarly, the corresponding ΔASA for class II pMHC complexes was $894.40 \pm 364.00\,\text{Å}^2$ (Ranganathan et al. 2008).

Intermolecular Hydrogen Bonds

The selectivity and stability of proteins and protein–protein complexes depend on various factors but one of the most important and major contribution comes from hydrogen bonds. Hydrogen bonds are the collective interaction of three atoms wherein a hydrogen atom is bound to donor electronegative atom and an acceptor electronegative atom in close proximity. The typical observed hydrogen bond distance is approximately 2.60–3.10 Å (1.00–1.20 Å between donor and hydrogen and 1.60–2.00 Å between hydrogen and acceptor). The significance of such bonding relies on both electronegative atoms being derived from the group: F, N, and O (Morrison and Boyd 1992). Only these elements are sufficiently negative and the hydrogen bound to any of these three elements is sufficiently positive for the required attraction to exist as their small atoms have a high concentration of negative charge on them. The interactions between side-chains are directed and their geometry is restricted due to the presence of hydrogen bonds. With decreasing bond length the strength of hydrogen bond usually increases.

Gap Volume

Gap volume is a measure of the volume enclosed by the two interacting molecular subunits. The gap volume between the MHC and the peptide in each complex can be computed using the SURFNET program (Laskowski 1991). This is done by the algorithm by placing a series of spheres (maximum radius 5.00 Å) between the surfaces of each pair of the MHC and the peptide subunit atoms, such that its surface is in contact with the surfaces of the atoms on either side. Upon interception by other atoms, the size of each sphere is reduced accordingly and is subsequently discarded if it falls below a minimum allowed radius (1.00 Å). The gap volume between the two subunits is calculated by taking into account the sizes of all the remaining allowable gap-spheres.

Gap Index

Electrostatic and geometric complementarity observed between associating molecules has always been an essential feature in receptor–ligand binding. Gap index (Jones and Thornton 1996) is a valuable method to evaluate complementarity of interacting interfaces:

$$\text{Gap index}\,(\mathring{A}) = \frac{\text{Gap volume between peptide and MHC}\,(\mathring{A}^3)}{\text{pMHC interface ASA}\,(\mathring{A}^2)} \tag{2}$$

The results for the mean gap indices of class I (0.95 ± 0.24 Å) and class II (1.12 ± 0.20 Å) pMHC complexes (Kangueane et al. 2001) indicate that the interacting surfaces in pMHC complexes are significantly complementary. The gap index in class II complexes is, on an average, higher than in class I complexes. This implies that the interface area of class I complexes is greater than its corresponding gap volume. On the contrary, in class II complexes, the gap volume is greater than the interface area. The complexes of different alleles in both class I and class II structures did not show much difference in their gap indices.

Supertype Classification Based on Structural Characteristics

Human leukocyte antigens (HLA) were until now classified based on the common structural features of HLA proteins (Doytchinova et al. 2004; Doytchinova and Flower 2005) and/or their functional binding specificities (Lund et al. 2004; Kangueane et al. 2005). These approaches leave the structural interaction characteristics among different HLA supertypes with antigenic peptides unexplored. We therefore classified 68 HLA class I molecules using the number of intermolecular hydrogen bonds between each HLA protein and its corresponding bound peptide, solvent accessibility of each pMHC complex, gap volume, and gap index as described above. This type of classification of the HLA proteins into supertypes helps in the identification of promiscuous T cell epitopes that bind multiple alleles and is thus the underlying reason for the development of successful epitope-based vaccines covering a wide number in the world population and all the ethnicities (Sette et al. 2001; Sette et al. 2002). The interaction parameters investigated in this study tend to vary among different alleles and were thus grouped in a supertype dependant manner. Our analysis resulted in successful classification of the HLA class I molecules into eight supertypes based on their crystallographic structures (Tong et al. 2007b). The HLA-A supertypes had three main clusters AI, AII, and AIII whereas HLA-B supertypes had five clusters namely BI, BII, BIII, BIV, and BV. Our data largely overlap the definition of binding motifs. The proposed methodology of classification which considers conformational information of both peptide and HLA proteins provides an alternative to the characterization of supertypes using either peptide or HLA protein information alone. A hierarchical clustering technique using the agglomerative algorithm (Doytchinova et al. 2004; Doytchinova and Flower 2005) was applied in this approach. The distance between the structures was computed by the single-linkage method based on the separation between the each pair of data points (Barnard and Downs 1992). The nearest neighbors were merged into clusters. Smaller clusters were then merged into larger clusters based on inter-cluster distances, until all structures are combined. We have considered the last three levels for defining HLA class I supertypes.

The MHC–Peptide Docking Protocol

Of all the techniques used for investigating intermolecular interactions, computer-simulated ligand binding or docking is the most powerful and widely used. The general purpose of docking simulation is: (i) to find the most probable translational, rotational, and conformational juxtaposition of a given ligand–receptor pair, and (ii) to evaluate the relative goodness-of-fit or how well a ligand can bind to the receptor. We now introduce a rapid and highly accurate docking protocol for the modeling of bound peptide ligands to the MHC receptor. We begin with the sequence of the ligand for which the structure is to be generated (peptide) and the availability of the target MHC receptor structure. Our docking protocol consists of four steps in all, out of which three essential steps are common for both class I and class II MHC complexes: (i) rigid docking of terminal residues of the peptide nonamer; (ii) loop closure of central residues by satisfaction of spatial constraints; (iii) followed by ab initio refinements of backbone and ligand interacting side-chain. However, for MHC class II complexes, the peptide length is usually between 12 and 15 residues and therefore there was a necessity to carry out addition of extra residues at both ends of the peptide nonamers as these residues usually extend out of the binding grove of the MHC molecule. Thus, step (iv) extension of flanking residues was included in the docking procedure for class II complexes. The flow diagram of the docking protocol is illustrated in Fig. 1.

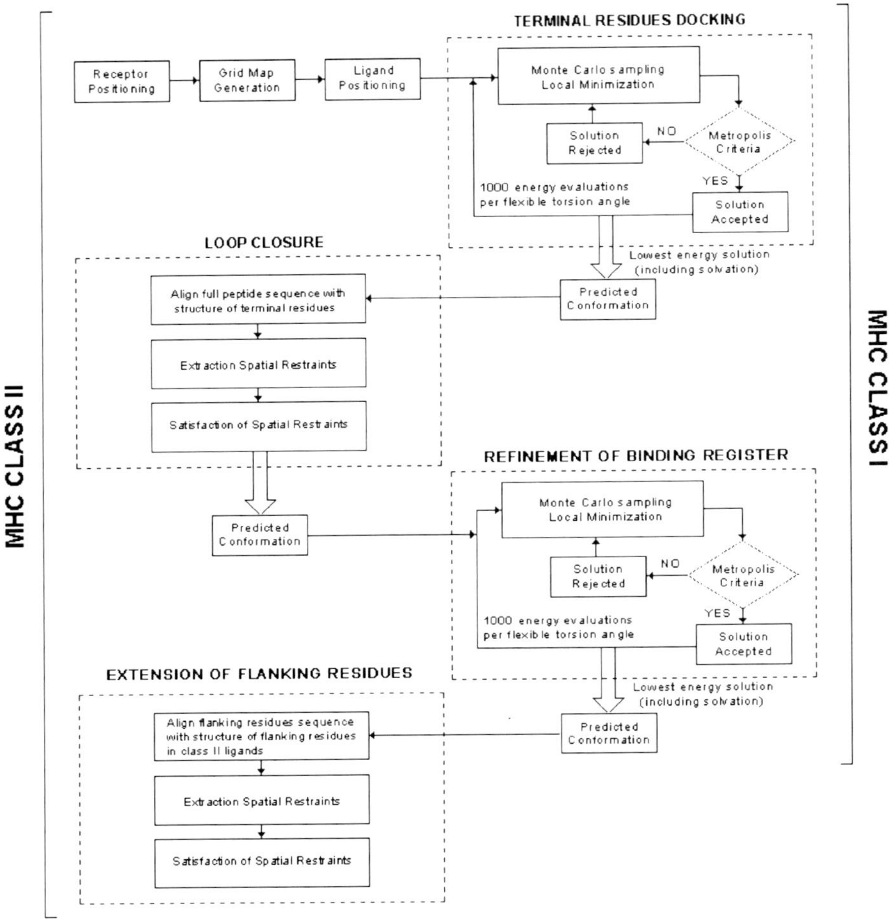

Fig. 1 Flowchart of the four-step docking procedure used in this chapter

Step 1: Rigid Docking of Nonamer Termini

To decide the number of combinations for two molecules within an enclosed sampling space is the key issue in docking simulation methodologies and an important factor that determines the best fit for the final output. There are six degrees of global-rotational and translational freedom of one molecule relative to the other, as well as one internal dihedral rotation per rotational bond. An increase in molecule size and sampling space increases search on the conformational space exponentially. Minimization of the conformational search space of ligand within the large sampling space enclosed by the MHC binding groove is a challenge in pMHC docking simulation. A possible approach to initiate docking simulations is to identify suitable anchor residues or nonamer termini (probes) for rigid docking. A probe must satisfy two criteria: (i) the anchor must have sufficient contact with the receptor, and (ii) the structure of the anchor must be highly conserved.

Peptide residues at the N and C termini of the nonamer are almost in invariant positions at the end of the binding groove of the MHC with mean backbone Cα RMSD within 0.15±0.14 Å (Tong et al. 2004) and are ideal for such purpose. To model each probe to the receptor, a fast soft-interaction energy function (Fernández-Recio et al. 2002) is adopted. This is performed using an internal coordinate mechanics (ICM) (Abagyan and Maxim 1999) global optimization algorithm, with flexible ligand interface side-chains and a grid map representation of the receptor energy localized to small cubic regions of 1.00 Å radius from the backbone of each probe. Within their respective grid map, each anchor residue performs a random walk. Using a Biased Monte Carlo procedure, which begins by pseudo-randomly selecting a set of torsion angles in the probe and subsequently finding the local energy minimum about those angles, the side-chain torsions were changed at random for each step. Upon satisfaction of the Metropolis criteria with probability min $(1, \exp[-\Delta G/RT])$, where R is the universal gas constant and T is the absolute temperature of the simulation, new conformations are adopted. To keep the positional variables of the ligand molecule close to the starting conformation, loose restraints were imposed on it. The stimulation temperature was set to 300 K. The internal energy of the probe and the intermolecular energy based on the same optimized potential maps used in the docking step together comprised the optimal energy function used during simulations:

$$E = E_{\mathrm{Hvw}} + E_{\mathrm{Cvw}} + 216E_{\mathrm{el}}^{\mathrm{solv}} + 253E_{\mathrm{hb}} + 435E_{\mathrm{hp}} + 0.20E_{\mathrm{solv}} \tag{3}$$

The internal energy included internal van der Waals interactions, hydrogen bonding, and torsion energy calculated with ECEPP/3 parameters, and the Coulomb electrostatic energy with a distance-dependent dielectric constant ($e = 4r$). In order to select the best-refined solutions the surface-based solvation energy and the configurational entropy of side-chains were included in the final energy.

Step 2: Loop Closure of Middle Residues

By satisfaction of spatial constraints (Sali and Blundell 1993) based on the allowed subspace for backbone dihedrals in accordance with the conformations of nonamer termini docked into the ends of the binding groove, an initial conformation of the central loop is generated at this stage. The three steps that are used to perform this are: (i) The alignment of the entire peptide sequence and the sequences of probes docked into the binding groove gives the distance and dihedral angle restraints on the peptide sequence. (ii) By extrapolation from the known 3D structures of probes in the alignment, expressed as probability density functions, the restraints on spatial features of the unknown central residues are derived. Stereo-chemical restraints include bond distances, bond angles, planarity of peptide groups and side-chain rings, chiralities of Cα atoms and side-chains, van der Waals contact distances and the bond lengths, bond angles, and dihedral angles of cysteine disulfide bridges.

(iii) Optimizing the molecular probability density function using variable target function technique that applies the conjugate gradients algorithm to positions of all non-hydrogen atoms satisfies spatial restraints on the unknown central residues.

Step 3: Refinement of Binding Register

Partial refinement was performed for both the ligand backbone and side-chain to improve the accuracy of the initial model using ICM Biased Monte Carlo procedure (Abagyan and Maxim 1999). By introducing partial flexibility to the ligand backbone, preliminary stages of refinements attempt to nullify the penalty derived from the initial rigid docking of terminal residues thus making this an effective and flexible docking procedure. Restraints were imposed upon the positional variables of the Cα atoms of probes to keep it close to the starting conformation. This refinement step is performed using the energy function:

$$E = E_{vw} + E_{hbonds} + E_{torsions} + E_{electr} + E_{solv} + E_{entropy} \quad (4)$$

Ligand and receptor side-chain torsions within 4.00 Å from the receptor were refined upon the final backbone structure.

Step 4: Extension of Flanking Residues

By now, MHC class I ligand models have already been fully constructed and this step is applicable only to MHC class II ligands. The only construction remaining is of the flanking residues that extend out of the MHC class II binding groove. The conformations of the flanking peptide residues are generated by satisfying the spatial constraints in the allowed subspace for backbone dihedrals defined by the conformation of the bound core nonameric peptide docked into the binding groove. This is again performed in three stages: (i) distance and dihedral angle restraints on the entire peptide sequence are derived from its alignment with the nonamer sequence in the binding groove; (ii) the restraints on spatial features of the flanking residues are derived by extrapolation from the known 3D structure of flanking residues in the alignment, expressed as probability density functions; and (iii) the spatial restraints on the flanking residues are then satisfied by optimization of the molecular probability density function using a variable target function technique that applies the conjugate gradients algorithm to positions of all non-hydrogen atoms.

Epitope Prediction

We will now compare the applications of this protocol for the discrimination of binders/non-binders from MHC class I and class II alleles. First, we discuss the docking of 68 peptides with known IC_{50} values and 12 peptides with experimental T cell proliferation values on to the binding groove of DQ3.2β MHC class II allele associated with several allergies and autoimmune diseases. Our model predicts DQ3.2β binding peptides with high accuracy [area under the receiver operating characteristic (ROC) curve $A_{ROC}>0.88$] (Tong et al. 2006b), compared with experimental data. Our investigation of the binding patterns of DQ3.2β peptides illustrates that several registers exist within a candidate binding peptide. Further analysis reveals that peptides with multiple registers occur predominantly for high-affinity binders (specificity = 0.95). We successfully predicted 20/23 (87%) binding registers with excellent discrimination of low-, medium-, and high-affinity binders. The results also proved that our method was of high precision with a sensitivity value as high as 0.81 (81%) for low-, medium-, and high-affinity binders.

In the second experiment we carried out docking of 51 DRB1*0402-specific desmoglein 3 (Dsg3) peptides with known IC_{50} values, 25 DRB1*0402-specific Dsg3 peptides with experimental T cell proliferation values and 6 DQB1*0503-specific Dsg3 peptides with experimental T cell proliferation values into the binding groove of Pemphigus vulgaris (PV) associated MHC class II alleles DRB1*0402 and DQB1*0503, respectively (Tong et al. 2006c). Docking of anchor peptide residues is performed using the docking procedure prior described followed by ab initio modeling of flanking residues. Our models present the best fit of each peptide into the binding cleft of each disease associated allele based on the following criteria: (i) pattern of hydrogen bonding to the MHC molecule; (ii) pattern of hydrophobic burial of peptide side-chains, and (iii) the absence of atomic clashes or repulsive contacts. Figure 2 illustrates the immunological hotspots that were predicted for both the alleles across the Dsg3 glycoprotein proteome. These immunological hotspots are the regions that contain immunogenic peptide epitopes that are highly suitable for vaccine design.

We were able to successfully predict all 25 and 5/6 peptides in test set II as high-binders for both DR and DQ alleles, respectively, with high accuracy ($A_{ROC}=0.93$) and specificity (SP = 0.80). These results were consistent with our earlier qualitative structural studies (Tong et al. 2006d). Furthermore, these results confirm that both DRB1*0402 and DQB1*0503 are strongly associated with PV. Our analysis revealed the existence of multiple immunodominant epitopes that may be responsible for both disease initiation and propagation in PV and also suggests that DRB1*0402 and DQB1*0503 may share similar specificities by binding peptides of different binding registers, thus providing a molecular mechanism for the dual HLA association observed in PV. In this example, pMHC residues were considered to be in contact if at least one pair of their non-hydrogen ("heavy") atoms was found to be within 4.00 Å radius (Fischer and Marquesee 2000). Intra-peptide interactions and intra-MHC interactions were not considered as they have minor influence on backbone structure. Any atom in the peptide and any atom in the MHC were considered to be experiencing atomic clash if their separation is below 2.00 Å (Samudrala and Moult 1997) for non-hydrogen atoms and below 1.60 Å for atoms participating in hydrogen bonds (Samanta et al. 2002; Wallace et al. 1995).

We also performed a docking and binding prediction study of the repertoires for HIV-1 p24gag and gp160gag glycoproteins that are known to bind the MHC class I allele HLA-Cw*0401 which plays a major role in the control of human immunodeficiency virus type 1 (HIV-1) infection. The analysis of predicted Cw*0401-

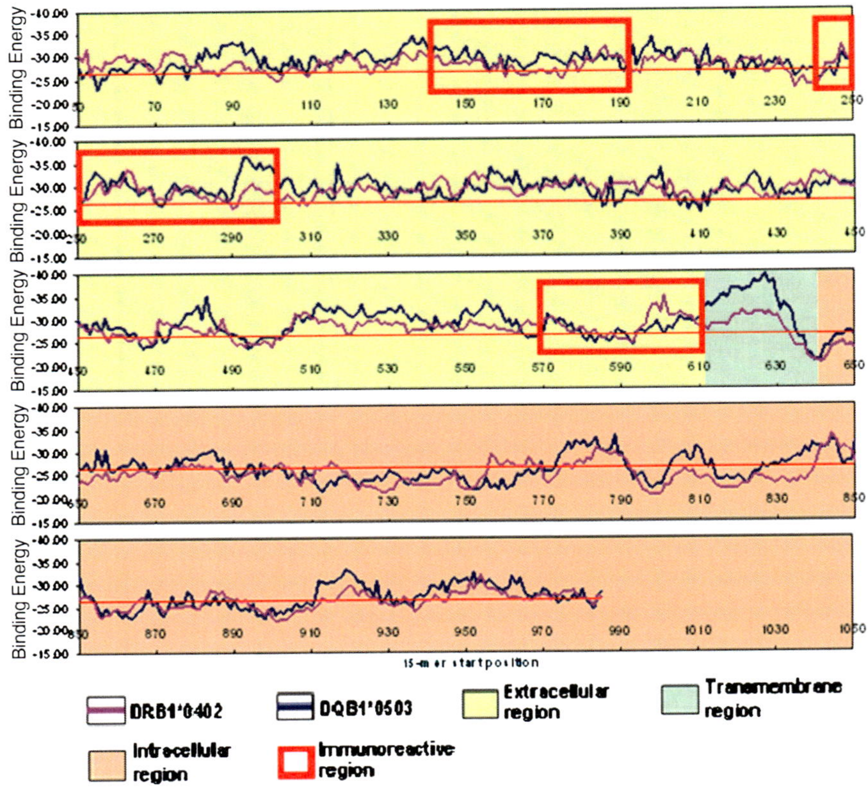

Fig. 2 Proteome-wide screening of Dsg3 peptides

binding peptides showed that anchor residues may not be restrictive and the Cw*0401 binding pockets may possibly accommodate a wide variety of peptides with common physico-chemical properties (Tong et al. 2007c). The potential Cw*0401-specific T cell epitopes are well distributed throughout both glycoproteins, with 14 and 9 immunological hotspots for HIV-1 p24gag and gp160gag glycoproteins, respectively. External validation results indicate that our Cw*0401 predictive model provides excellent discrimination between binding and the non-binding ligands with high accuracy ($A_{ROC}=0.93$) and a sensitivity of 76% (SE = 0.76) and a specificity as high as 95% (SP = 0.95). Our results strongly indicate that Cw*0401 can bind antigenic peptides in amounts comparable to both HLA-A and -B molecules, and support the existence of a potentially large number of Cw*0401-specific T cell epitopes.

Table 1 gives a summary of the overall comparison of the results from the three experiments described above. An important thing to note is the diminishing training datasets used in the three experiments which suggests that our prediction model can

Table 1 Comparison of the training set, test set, and results of the three prediction experiments

DQ3.2β (Tong et al. 2006b)	Training set	56 binding and 30 non-binding conformations from experimentally determined binding and non-binding peptides
	Test set	I – 68 peptides with known IC_{50} values II – 12 peptides with known T cell proliferation values
	Results	I – A_{ROC}=0.88, SE=0. 81, and SP=0.95. 20/23 (87%) binding registers were predicted correctly. II – Top five predictions (SE=0.95) have known T cell proliferation values
DR and DQ (Tong et al. 2006c)	Training set	8 DRB1*0402-specific Dsg3 peptides 8 DQB1*0503-specific Dsg3 peptides
	Test set	I – 51 DR-specific Dsg3 peptides with known IC_{50} values II – 25 DR and 6 DQ-specific Dsg3 peptides with experimental T cell proliferation values
	Results	I – A_{ROC}=0.93, SE=0.70 and SP=0.95. II – All 25 DR-specific peptides in test set II were predicted as high binders (SE=0.65, SP=0.80). 5/6 DQ-specific peptides (all true positives) were determined with a cutoff of –26.64 kJ/mol
HLA-Cw*0401 (Tong et al. 2007c)	Training set	6 peptide sequences with known IC_{50} values
	Test set	58 peptides known to bind Cw*0401
	Results	Test set accuracy: A_{ROC}=0.93, SE=0.76, and SP=0.95. 14 and 9 potential Cw*0401-specific T cell immunogenic regions or hotspots were predicted for HIV-1 p24gag and gp160gag glycoproteins, respectively

give excellent results with as low as six peptides in the training dataset. It, however, is yet to be determined as to what proportion of these predicted peptides may be expressed at the cell surface and are capable of eliciting functional T cell responses. One aspect of this would be to look at the TR/pMHC complex on the whole.

TR/pMHC Interaction

TR/pMHC interaction is the most essential binding step in the entire adaptive immune response cascade. It is this interaction that is responsible for eliciting a T cell response. After an endogenous or exogenous peptide is bound to the MHC class I or class II, respectively, the pMHC complex is transported to the membrane of the antigen presenting cell (APC) and is presented at the cell surface for surveillance by the TR which then binds to the pMHC and forms the TR/pMHC complex. This binding is also partly determined by the cluster of differentiation (CD) molecules present on the membrane of the T cells as these bind specifically to MHC class I and class II proteins. Therefore, class I pMHC molecules stimulate CD8+ cytotoxic T cells which directly kill the infected cells whereas class II pMHC molecules stimulate CD4+ helper T cells which in turn activate B cells, leading to antibody production. Three-dimensional structures of the pMHC complex and the TR are essential and play a vital role in the activation of the Adaptive immune system. With the chance of 1 in 2,000 antigenic peptides being able to stimulate T cells (Yewdell and Bennink 1999), finding immunogenic peptide epitopes poses a great challenge. In view of this enormity, it is experimentally impossible to scan all the putative peptides arising from the proteomes of all pathogens known today. An in-depth analysis of TR/pMHC complexes using structural immunoinformatics combines the power of computational analysis with detailed structural data to accelerate immune system research and provides clues for the development of vaccines for immunotherapeutic applications.

Figure 3 depicts the TR footprint on a pMHC complex. The residues (both MHC and peptide residues) that comprise the footprint are the key to pMHC recognition by the TR molecule via the variable regions of the α and β chains. The residues on the TR molecule that interact with the residues of the pMHC complex (Fig. 4) are also equally essential to this process as they recognize the corresponding residues on the pMHC complex and form H-bonds with them to anchor the TR molecule to the pMHC complex.

However, it is the peptide epitope that is important for vaccine development as a few of its residues take part in the peptide–MHC binding and the others take part in the TR/pMHC recognition and binding. Therefore, it can be inferred that it is the epitope that acts as a key to unlock the immune cascade and thereby plays the most important role in this major defense mechanism in higher vertebrates.

A detailed analysis of the types of inter-residue interactions observed in this complex is shown in Fig. 5. The peptide residues, Gly4(C), Thr8(C), and Val6(C), interact with TR residues, Gln52(E), Asp32(E), Gln52(E), and Ser99(E), respectively, to form H-bonds and thereby stabilize the TR/pMHC complex besides anchoring TR to the pMHC complex. Gly97(D) and Ile53(E) of the TR are also

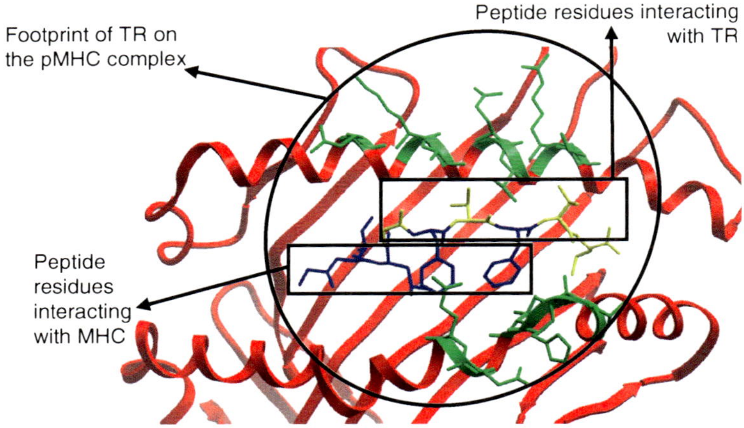

Fig. 3 A schematic structure of the top view of the MHC groove and the bound peptide in a class I TR/pMHC complex (PDB ID – 1OGA) (Stewart-Jones et al. 2003) in Cα trace ribbon representation, with MHC in red and peptide in blue. Residues interacting with TR are highlighted in green (MHC) and yellow (peptide), with heavy side-chain atoms shown in stick representation. The black oval contains these residues or the footprint of the TR on the pMHC complex

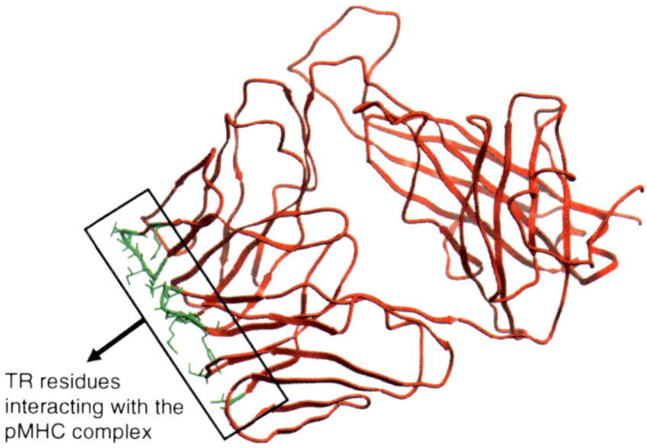

Fig. 4 Schematic Cα trace ribbon representation of the T cell receptor (PDB ID – 1OGA) in red, with the residues in its variable regions interacting with the pMHC complex highlighted in green

involved in hydrophobic interactions with Gly4(C) and Phe5(C), Thr8(C), respectively, thereby contributing to the overall stability of the TR/pMHC complex. Such interactions are also seen between the MHC and the peptide residues suggesting the importance of peptide epitopes for immunotherapy and vaccine development. Water bridges (not shown) also play a significant role in this TR/pMHC complex formation and stability.

Analysis of the 1OGA Complex

We now extend our interaction parameters to the TR level in order to analyze the 1OGA TR/pMHC structure, using the formulae described earlier. The interface area for the TR/pMHC complex is 733.55 $Å^2$ which is lower compared to that of the pMHC complex (856.00 $Å^2$) for the same structure. This suggests that the TR/pMHC binding is localized. As seen in Fig. 5 , the peptide is sandwiched between the TR and the pMHC and is fairly well buried in terms of the accessible surface area.

Figure 5 indicates the presence of four H-bonds between the peptide and the TR residues [Gly4(C) with Gln52(E), Val6(C) with Gln52(E), and Ser99(E), Thr8(C) with Asp32(E)]. The average H-bond length between the peptide and the TR residues is 2.95 Å. H-bonding (not shown) between the TR residue, Arg98(E) and the MHC residues, Ala150(A) and Gln155(A) with an average bond length of 2.92 Å, also plays a vital role in anchoring the TR onto the pMHC complex. Notably, 12 hydrogen bonds are seen to stabilize and anchor the peptide firmly onto the MHC molecule, of length 2.67–3.08 Å. Hydrophobic interactions are also seen to occur specially between the MHC and the TR molecules, involving Val152(A), Gln155(A), and Lys66(A) residues of MHC (interacting TR residues not shown in Fig. 5), while Thr73(A) interacts strongly with its neighboring TR residue, Ile53(E).

The gap volume between pMHC and TR was calculated using a molecular model of only the interacting regions from both pMHC and TR, generated from the

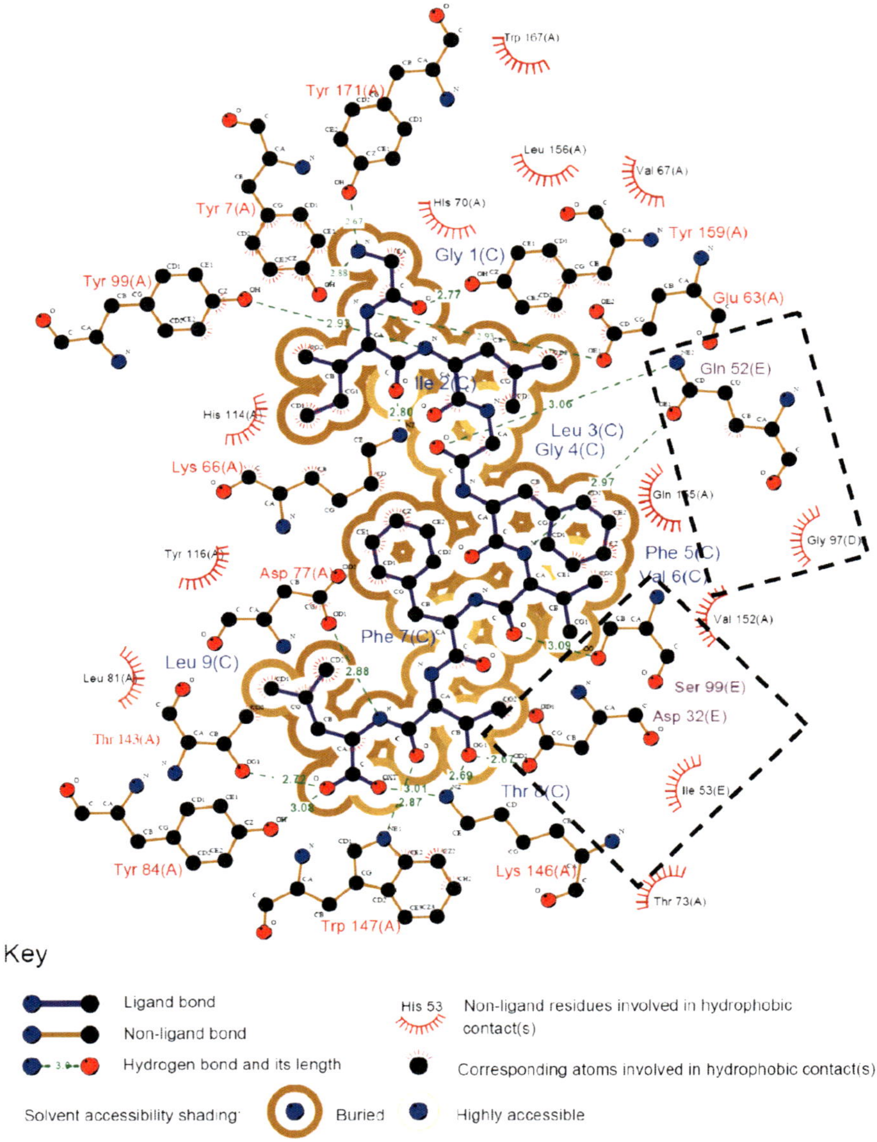

Fig. 5 A LIGPLOT schematic diagram of the peptide ligand from the three dimensional structure of class I TR/pMHC complex (PDB ID – 1OGA) showing the solvent accessibility and the interactions between its residues and the corresponding MHC and TR residues. The TR residues, Gln52(E), Asp32(E), Ser99(E), Ile53(E), and Gly97(D) that together anchor the TR on the pMHC are shown in dotted rectangles. The letters in the brackets correspond to the respective chain IDs to which the residues belong. (A) – MHC α chain, (C) – peptide, (D) and (E) – TR α and β chains, respectively

template structure, 1OGA using MODELLER (Sali and Blundell 1993). The gap volume was then computed to be 3304.43 Å^3 using SURFNET (Laskowski 1991), depicted by the blue region in Fig. 6. This value of the gap volume between the TR and pMHC is very large compared to that between the peptide and MHC suggesting that the binding between peptide and MHC is much stronger than that between

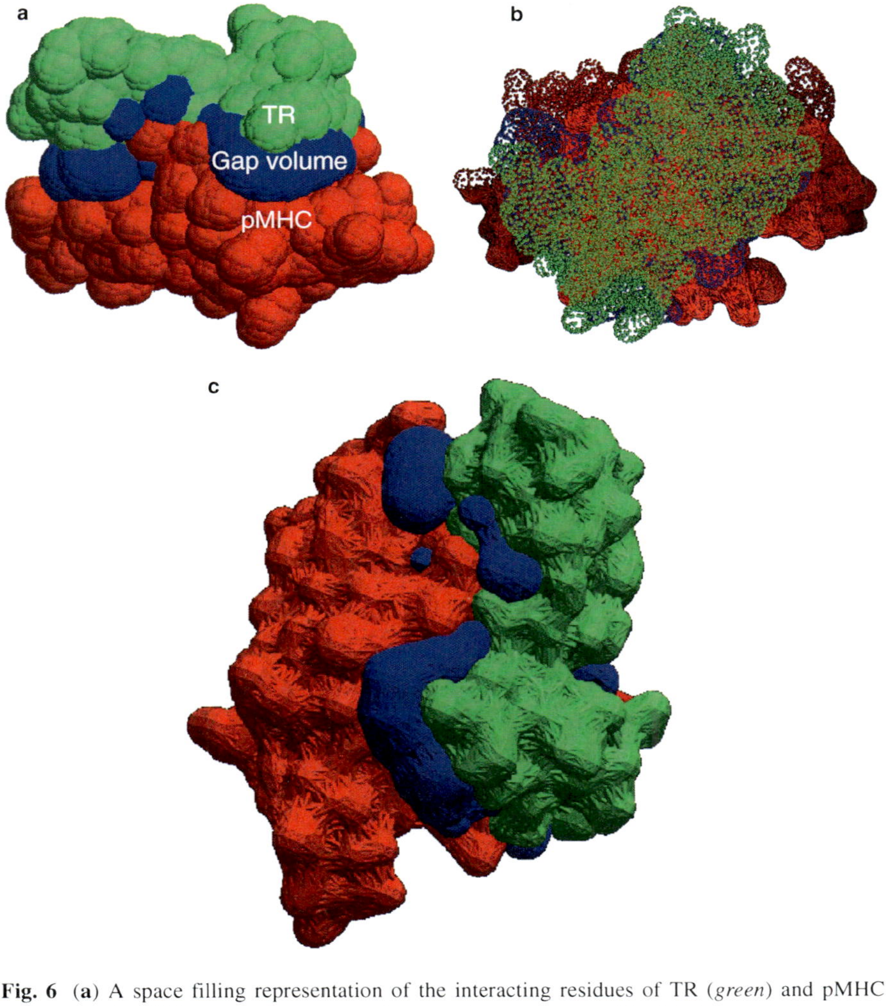

Fig. 6 (**a**) A space filling representation of the interacting residues of TR (*green*) and pMHC (*red*) from the 1OGA crystal structure, showing the gap volume (*blue*) between the two complexes. A fairly large gap volume supports the theory that the binding is not very strong, unlike the pMHC binding. The surface representation file used for visualization was generated using SURFNET (Laskowski 1991). (**b**) Top view of the TR/pMHC complex shown in an atomic mesh representation. The TR binds to the pMHC at an angle of 69° (Stewart-Jones et al. 2003). (**c**) Side view of the TR/pMHC interacting regions, in ball and stick surface representation

pMHC and TR. These results underline the importance of strong peptide binders in the first step of the entire adaptive immune response cascade. A large gap volume and a small interface area indicate that the gap index of the TR/pMHC complex is high (4.50 Å), compared to that of the peptide and the MHC (0.60 Å). This suggests that the electrostatic and the geometric complementarities of the TR and pMHC are not as significant as between the peptide and MHC.

Conclusion

Our analysis and extensive studies on peptide–MHC interactions have revealed structural features that can be analyzed in terms of the parameters governing the pMHC complex formation. We have now extended this formalism to defining the interaction between TR and pMHC, relevant for immune system activation. Based on our pMHC analyses, we have developed methods to successfully predict T cell epitopes in accordance with their MHC binding specificities. The next challenge is to extend this methodology to the unexplored TR level as this would greatly improve the efficacy of our prediction model, in separating a large number of predicted MHC-binding peptides from true T cell epitopes. The complexities involved in methodology development and the computational costs incurred in docking peptides and proteins have hindered the progress of structure-based prediction techniques. In the era of high throughput and distributed computing over global grids, the necessary computational requirements for large-scale structure-based screening of potential T cell epitopes are now available. We can therefore expect new structure-based approaches to predicting promiscuous peptide epitopes for MHC supertypes and TR activation, for the design of sub-type-specific vaccines with wide population coverage. Large-scale structure-based screening helps overcome the barriers of insufficient training data and the lack of peptide binding motifs, especially for MHC class II alleles by cutting down the lead time involved in experimental vaccine development methods, resulting in the production of effective and highly specific peptide vaccines.

References

Abagyan R, Maxim T (1999) Ab Initio Folding of Peptides by the Optimal-Bias Monte Carlo Minimization Procedure. J Comput Phys 151:402–421

Barnard JM, Downs GM (1992) Clustering of chemical structures on the basis of two-dimensional similarity measures. J Chem Inf Comput Sci 32:644–649

Berman HM, Westbrook J, Feng Z et al (2000) The protein data bank. Nucleic Acids Res 28:235–242

Chothia C, Janin J (1975) Principles of protein–protein recognition. Nature 28:705–708

Doytchinova IA, Flower DR (2005) *In silico* identification of supertypes for class II MHCs. J Immunol 174:7085–7095

Doytchinova IA, Guan P, Flower DR (2004) Identifying human MHC supertypes using bioinformatic methods. J Immunol 172:4314–4323

Fernández-Recio J, Totrov M, Abagyan R (2002) Soft protein–protein docking in internal coordinates. Protein Sci 11:280–291
Fischer KF, Marquesee S (2000) A rapid test for identification of autonomous folding units in proteins. J Mol Biol 302:701–712
Govindarajan KR, Kangueane P, Tan TW et al (2003) MPID: MHC-Peptide Interaction Database for sequence-structure-function information on peptides binding to MHC molecules. Bioinformatics 19:309–310
Hubbard SJ, Thornton JM (1993) "NACCESS" computer Program, Department of Biochemistry and Molecular Biology, University College, London
Jones S, Thornton JM (1996) Principles of protein–protein interactions. Proc Natl Acad Sci 93:13–20
Kangueane P, Sakharkar MK, Kolatkar PR et al (2001) Towards the MHC-peptide combinatorics. Hum Immunol 62:539–556
Kangueane P, Sakharkar MK, Rajaseger G et al (2005) A framework to sub-type HLA supertypes. Front Biosci 10:879–886
Laskowski RA (1991) "SURFNET" computer program, Department of Biochemistry and Molecular Biology, University College, London
Lund O, Nielsen M, Kesmir C et al (2004) Definition of supertypes for HLA molecules using clustering of specificity matrices. Immunogenetics 55:797–810
Morrison RT, Boyd RN (1992) Organic chemistry, vol 6. Prentice Hall, USA
Ranganathan S, Tong JC, Tan TW (2008) Structural immunoinformatics. In: Schonbach C, Ranganathan S, Brusic V (eds) Springer, Immunomics Reviews Series, chap. 3, pp 51–61
Sali A, Blundell TL (1993) Comparative protein modeling by satisfaction of spatial restraints. J Mol Biol 234:774–815
Samanta U, Bahadur RP, Chakrabarti P (2002) Quantifying the accessible surface area of protein residues in their local environment. Prot Eng 15:659–667
Samudrala R, Moult J (1997) Handling context-sensitivity in protein structures using graph theory: Bona fide prediction. Proteins 1(Suppl):43–49
Sette A, Livingston B, McKinney D et al (2001) The development of multi-epitope vaccines: Epitope identification, vaccine design and clinical evaluation. Biologicals 29:271–276
Sette A, Newman M, Livingston B et al (2002) Optimizing vaccine design for cellular processing, MHC binding and TCR recognition. Tissue Antigens 59:443–451
Stewart-Jones GB, McMichael AJ, Bell JI et al (2003) A structural basis for immunodominant human T cell receptor recognition. Nat Immunol 4:657–663
Tong JC, Bramson J, Kanduc D et al (2006a) Modeling the bound conformation of pemphigus vulgaris associated peptides to MHC class II DR and DQ alleles. Immunome Res 2:1
Tong JC, Kong L, Tan TW et al (2006b) MPID-T: Database for sequence-structure-function information on TCR-peptide-MHC interactions. Appl Bioinform 5:111–114
Tong JC, Tan TW, Ranganathan S (2004) Modeling the structure of bound peptide ligands to major histocompatibility complex. Protein Sci 13:2523–2532
Tong JC, Tan TW, Sinha AA et al (2006c) Prediction of desmoglein-3 peptides reveals multiple shared T-cell epitopes in HLA DR4- and DR6-associated pemphigus vulgaris. BMC Bioinform 18:7
Tong JC, Tan TW, Ranganathan S (2007a) Methods and protocols for prediction of immunogenic epitopes. Brief Bioinform 8:96–108
Tong JC, Tan TW, Ranganathan S (2007b) *In silico* grouping of peptide/HLA class I complexes using structural interaction characteristics. Bioinformatics 23:177–183
Tong JC, Zhang GL, August JT et al (2006d) Prediction of HLA-DQ3.2β Ligands: Evidence of multiple registers in class II binding peptides. Bioinformatics 22:1232–1238
Tong JC, Zhang GL, August JT et al (2007c) *In silico* characterization of immunogenic epitopes presented by HLA-Cw*0401. Immunome Res 3:7
Wallace AC, Laskowski RA, Thornton JM (1995) LIGPLOT: A program to generate schematic diagrams of protein–ligand interactions. Prot Eng 8:127–134
Yewdell JW, Bennink JR (1999) Immunodominance in major histocompatibility complex class I-restricted T lymphocyte responses. Annu Rev Immunol 17:51–88

Discovery of Conserved Epitopes Through Sequence Variability Analyses

Carmen M. Díez-Rivero and Pedro Reche

Introduction

Many pathogens exhibit high mutation rates, generating new genetic variants that are resistant to an existing immune response to earlier pathogen subtypes (Mendis et al. 1991; Phillips et al. 1991; Weber and Elliott 2002), difficulting the task of vaccine development. It is therefore important to focus on conserved regions during the process of vaccine design.

Several research groups have tried to develop vaccines based on quimeric consensus sequences (Thomsona et al. 2005). However, these vaccines have a major disadvantage as chimeric consensus proteins still bear nonconserved connecting regions, which might be more inmunogenic than conserved ones and thus truncate the development of a protective immune response. Nonprotective immunodominance can however be overcome using antigenic determinants (epitopes) as vaccines, as one can drive the immune response only towards the conserved epitopes of interest (Sette et al. 2002; Tsuji and Zavala 2001; Disis et al. 2001; Reche et al. 2006).

The estimation of sequence variability from MSAs of protein antigens also provides a means to identify conserved antigenic determinants. In this chapter, we will illustrate the use of PVS (García-Boronat et al. 2008), a Protein Variability Server that has been tuned to facilitate the discovery of conserved epitopes. Specifically, we will use PVS to obtain the conserved regions of the HIV-1 gp120 and gp41 proteins, identifying those that are solvent exposed, and therefore, likely the targets of cross-neutralizing antibodies (Abs). Likewise, we will use PVS to generate a variability-masked sequence of the HIV-1 gp120 protein, which will be targeted for T cell epitope predictions. Epitope-vaccine development requires confirming the immunogenicity of vaccine candidates, which consumes a vast amount of time and resources. Interestingly, sequence variability analyses in PVS dramatically reduce the number of potential epitope-vaccine candidates one would need to consider. PVS is freely available at the site http://imed.med.ucm.es/PVS.

P. Reche (✉)
Facultad de Medicina, Departamento de Immunología (Microbiología I), Universidad Complutense de Madrid, Pabellón 5°, planta 4ª, 28040, Madrid, Spain
e-mail: parecheg@med.umc.es

D.R. Flower et al. (eds.), *Bioinformatics for Immunomics*, Immunomics Reviews 3,
DOI 10.1007/978-1-4419-0540-6_8,

Materials and Methods

MSAs

For this study two proteins are used: The gp120 (residues 31-183 in gp160) and the gp41 (residues 528–674 in gp160), which are both membrane glycoproteins of HIV-1 (strain H2XB2). Both the gp120 and gp41 MSAs, were generated from 359 representative sequences of the HIV-1 clades A (73), B (85), C (85), D (51) and 01_AE (65) using the program MUSCLE (Edgar 2004). The gp41 and gp120 MSAs are available at http://imed.med.ucm.es/PVS/supplemental/gp120_pvs.html and http://imed.med.ucm.es/PVS/supplemental/gp41_pvs.html, respectively.

PVS Description and Usage

PVS (Protein Variability Server) is a web-based tool (Fig. 1) that following a protein sequence variability analysis performs several tasks that are relevant for structure-

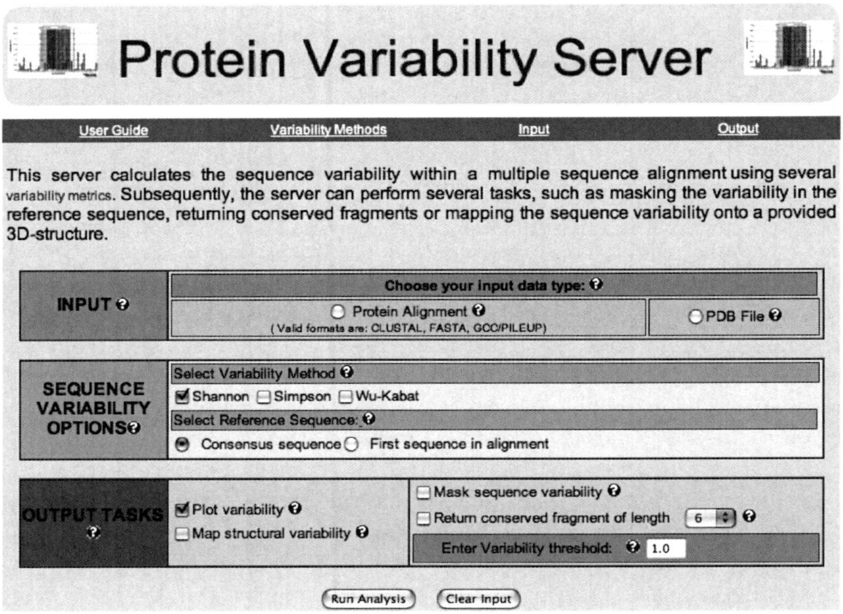

Fig. 1 *PVS web interface.* The web interface is divided into the INPUT, SEQUENCE VARIABILITY OPTIONS and OUTPUT TASKS sections which overall facilitate an intuitive use of the server. The web interface also provides links to help pages and specific information regarding the elements featured by the server accessible from the question mark icons

function studies and vaccine design. PVS main input is an MSA provided by the user, but it can also take a PDB file as main input, generating an MSA from it (for details see García-Boronat et al. 2008) The sequence variability in the MSA is computed *per site* using three different metrics: The Shannon Diversity index (Shannon Entropy) (Shannon 1948), the Simpson Diversity Index (Simpson 1949) and the Wu-Kabat Variability Coefficient (Wu and Kabat 1970). In this study, we have selected the Shannon Diversity Index (H) as the variability metric. H ranges from 0 (only one amino acid type is present at that position) to 4.322 (all 20 amino acids are equally represented in that position). Note, that for a site including gaps the maximum value of H will be 4.39. A site with a value of H under 1.0 is indicative of a site with very low variability (Reche and Reinherz 2003).

PVS optional tasks include that of plotting the variability in MSA – computed for each selected variability method – against a sequence consisting of a consensus sequence or the first sequence in the MSA. If the task "map structure variability" is selected and a PDB with relevant 3D-coordinates is submitted, PVS will map the sequence variability in the MSA onto the provided 3D-structure. Mapping the sequence variability onto the provided PDB is achieved by simply replacing the B-factor of the relevant residues with the variability values. Variability mapped 3D-structures can be visualized and manipulated interactively using JMOL (http://jmol.sourceforge.net/). The variability is shown in the 3D-structrure using a color scale that goes from blue for constant residues to red for highly variable residues. PVS also offers the possibility of returning the "conserved fragments." A variability threshold (*Vt*) and a minimum length of the conserved fragments need to be provided with this option. Under these selections, if a PDB is provided, PVS will also display a graph of the protein sequence with the conserved fragments shown in blue. By clicking on a fragment, one can locate the fragment on the 3D structure.

Finally, PVS can return the selected reference sequence with the variable positions masked. Specifically, those residues with variability greater than a user selected threshold will be shown with a "." symbol. The returned masked sequence is in FASTA format and can be directly submitted to RANKPEP (Reche and Reinherz 2007; Reche et al. 2004; Reche et al. 2002), a T cell epitope prediction tool that can anticipate conserved T-cell epitopes from a variability-masked sequence.

Results and Conclusion

Sequence variability is exploited by biological systems to generate functional heterogeneity (e.g., receptors involved in antigen recognition). Therefore, sequence variability analyses have traditionally been used to fill gaps in structural knowledge (Wu and Kabat 1970; Reche and Reinherz 2003). In addition, sequence variability analyses are also important for vaccine development as they also enable the identification of conserved antigenic determinants (Reche et al. 2006). For that purpose, we recently developed PVS, a web-based tool for protein variability analysis,

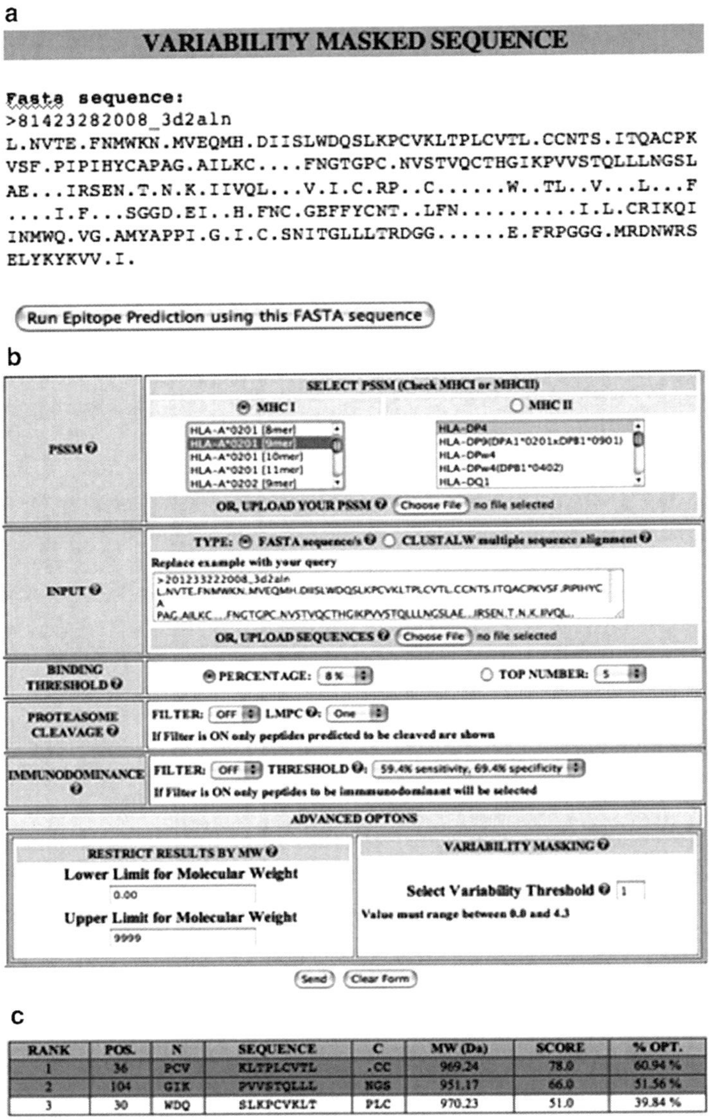

Fig. 2 *PVS and T cell epitope predictions.* (**a**) *Variability-masked sequence*. The shown sequence obtained from an MSA of HIV-1 gp120 (consensus sequence was selected as the reference sequence). The sequence is in FASTA format and positions indicated by dots, ".", display a variability > 1.0. (**b**) *Rankpep web interface.* By clicking on the button "Run Epitope Predictions" one will directly submit this sequence for conserved T cell epitope predictions *using* the RANKPEP algorithm. (**c**) RANKPEP results for the variability-masked sequence of the gp120. Only fragments KLTPLCUTL and PVVSTQLLL were predicted to have a binding score above the threshold

which implements several features that are thought to facilitate epitope-vaccine design. Next we will discuss such features using HIV-1 as the pathogenic model.

PVS can be used to facilitate the identification of conserved T cell epitopes. As an example we used an MSA from the HIV-1 gp120 protein (see Sect. 1 for details) to first obtain a variability masked sequence (Fig. 2a), which was subsequently targeted for the prediction of CD8+ T cell epitopes restricted by the HLA I molecule A*0201 (Fig. 2b). Interestingly, only two T cell epitopes (KLTPLCVTL and PVVSTQLLL) were predicted to have a binding score above the threshold (Fig. 2c) In comparison, the complete gp120 sequence (strain H2XB2) would yield 10 different epitopes. Thus, regardless of the predictive power of RANKPEP, this strategy saves the time, effort and resources that one will need to confirm non-conserved T cell epitopes that are not as suitable for epitope-vaccine design.

PVS results can also be useful for the identification of conserved B cell epitopes, the antigenic determinants of Abs. For example, the ectodomain of HIV-1 gp41 is known to be the target of various broadly neutralizing Abs (Zolla-Pazner 2004). When PVS is run with an MSA of this protein, 7 highly conserved fragments of 6 of more residues are returned (Table 1). Interestingly, fragments WGCSGK and WLWYIK encompass the antigenic determinants of the monoclonal Abs CL3 and ZE10, both broadly neutralizing. As we can see, the targets of broadly neutralizing Abs lie within conserved fragments.

Abs only recognize solvent-exposed epitopes, and most of them are conformational –although, some can also be linear–. To help identifying solvent-exposed fragments, PVS also allows exploring the location of the conserved fragments in the 3D-structure of the protein (when available). The use of such solvent-exposed conserved fragments as immunogens greatly increases the chance of raising Abs that are both, crossreactive with the native antigen and broadly neutralizing. For example, Table 2 shows that there are only eight highly conserved fragments lying within the reported gp120 structure (PDB 1RZK, chain G).

However, by mapping the conserved gp120 fragments onto the 3D-structure (Fig. 3) one could see that only fragment 2 and fragment 3 and significant portions of fragments 1, 4 and 6 are accessible to the solvent. Therefore, these solvent-exposed

Table 1 Conserved fragments in the ectodomain of HIV-1 gp41 calculated by PVS

N	Start	End	Sequence
1	1	7	S T M G A A S
2	9	25	T L T V Q A R Q L L S G I V Q Q Q
3	27	55	N L L R A I E A Q Q H L L Q L T V W G I K Q L Q A R V L A
4	62	67	D Q Q L L G
5	69	74	W G C S G K
6	87	92	S W S N K S
7	153	158	W L W Y I K

Fragments were selected to have six or more consecutive residues with $H \leq 1$, and were obtained form an MSA of the HIV-1 gp41 ectodomain

Table 2 Conserved fragments of the HIV-1 glycoprotein gp120 calculated by PVS

N	Start	End	Sequence
1	22	44	D I I S L W D Q S L K P C V K L T P L C V T L
2	52	61	I T Q A C P K V S F
3	63	73	P I P I H Y C A P A G
4	93	119	N V S T V Q C T H G I K P V V S T Q L L L N G S L A E
5	202	209	G E F F Y C N T
6	232	242	C R I K Q I I N M W Q
7	261	273	S N I T G L L L T R D G G
8	289	303	M R D N W R S E L Y K Y K V V

Fragments were selected to have eight or more consecutive residues with $H \leq 1$, and were obtained from an MSA of HIV-1 gp120 (See Material and Methods). The "Map structure variability" task was selected and chain G of PDB 1RZK containing the 3D-coordinates of HIV-1 gp120 was entered in the server. Relevant sequence in PDB is considerably shorter than that of MSA, and only those fragments mapping within the PDB sequence are reported by the server

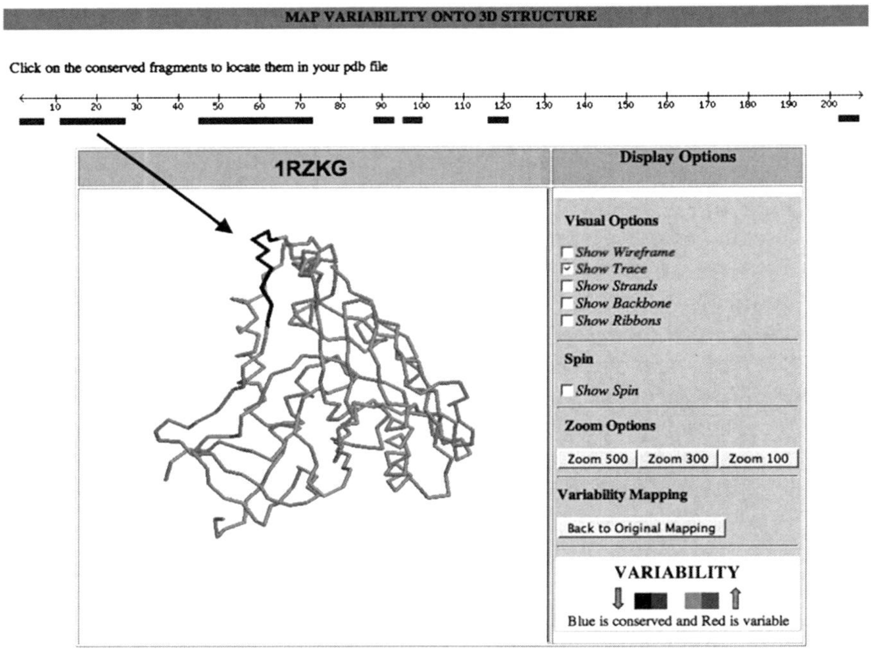

Fig. 3 *Exploring solvent accessibility of conserved fragments.* Arrow shows the location of fragment 2 (ITQACPKVSF) in the 3D-structure of gp120 (chain G of PDB 1RZK). It was located on the 3D-structure by simply clicking on the corresponding fragment shown under the linear representation of gp120

fragments are the only peptides from HIV-1 gp120 that may elicit both cross-neutralizing cross-reactive Abs with the native gp120.

Acknowledgments The work and the authors were supported by grant SAF2006-07879 from the "Ministerio de Educación y Ciencia" of Spain to PR.

References

Disis ML, Knutson KL, McNeel DG et al (2001) Clinical translation of peptide-based vaccine trials: The HER-2/neumodel. Crit Rev Immunol 21:263–274

Edgar RC (2004) MUSCLE: Multiple sequence alignment with high accuracy and high throughput. Nucleic Acids Res 32:1792–1797

García-Boronat M, Diez-Rivero CM, Reinherz EL et al (2008) PVS: A web server for protein sequence variability analysis tuned to facilitate conserved epitope discovery. Nucleic Acids Res 36:W35–W41

Mendis KN, David PH, Carter R (1991) Antigenic polymorphism in malaria: Is it an important mechanism for immune evasion? Immunol Today 12:A34–A37

Phillips RE, Rowland-Jones S et al (1991) Human immunodeficiency virus genetic variation that can escape cytotoxic T cell recognition. Nature 354:453–459

Reche PA, Reinherz EL (2003) Sequence variability analysis of human class I and class II MHC molecules: Functional and structural correlates of amino acid polymorphisms. J Mol Biol 331:623–641

Reche PA, Reinherz EL (2007) Prediction of peptide-MHC binding using profiles. Mol Biol 409:185–200

Reche PA, Glutting JP, Reinherz EL (2002) Prediction of MHC class I binding peptides using profile motifs. Hum Immunol 63:701–709

Reche PA, Glutting J-P, Reinherz EL (2004) Enhancement to the RANKPEP resource for the prediction of peptide binding to MHC molecules using profiles. Immunogenetics 56:405–419

Reche PA, Keskin DB, Hussey RE et al (2006) Elicitation from virus-naive individuals of cytotoxic T lymphocytes directed against conserved HIV-1 epitopes. Med Immunol 5:1

Sette A, Newman M, Livingston B et al (2002) Optimizing vaccine design for cellular processing, MHC binding and TCR recognition. Tissue Antigens 59:443–451

Shannon CE (1948) The mathematical theory of communication. Bell Syst Tech J 27(379–423): 623–656

Simpson EH (1949) Measurement of diversity. Nature 163:688

Thomsona SA, Jaramillo AB, Shoobridge M et al (2005) Development of a synthetic consensus sequence scrambled antigen HIV-1 vaccine designed for global use. Vaccine 23:4647–4657

Tsuji M, Zavala F (2001) Peptide-based subunit vaccines against preerythrocytic stages of malaria parasites. Mol Immunol 38:433–442

Weber F, Elliott RM (2002) Antigenic drift, antigenic shift and interferon antagonists: How bunyaviruses counteract the immune system. Virus Res 88:129–136

Wu TT, Kabat EA (1970) An analysis of the sequences of the variable regions of Bence Jones proteins and myeloma light chains and their implications for antibody complementarity. J Exp Med 132:211–250

Zolla-Pazner S (2004) Identifying epitopes of HIV-1 that induce protective antibodies. Nat Rev Immunol 4:199–210

Tunable Detectors for Artificial Immune Systems: From Model to Algorithm

Paul S. Andrews and Jon Timmis

Introduction

Artificial immune systems (AIS) combine elements of immunology with the engineering sciences, and are typically defined as "adaptive systems, inspired by theoretical immunology and observed immune functions, principles and models, which are applied to problem solving" (de Castro and Timmis 2002). A criticism of biologically-inspired engineering approaches such as AIS is that their development typically suffers from "reasoning by metaphor" (Stepney et al. 2005), proceeding directly from naive biological models to algorithms, with little investigation of the relevant biological properties. To tackle this, the conceptual framework approach of Stepney et al. (Stepney et al. 2005) aims to facilitate the development of bio-inspired algorithms in a more principled way than has been previously observed. It suggests that biologicallyinspired algorithms are designed through a series of observational and modelling stages in order to identify the key characteristics of the immunological process on which the AIS will be based. By building abstract models of the biology, investigations can be carried out that are not available to the actual biological system. The insight gained from these models should then lead to the construction of the bio-inspired algorithms and frameworks. It is suggested that algorithms developed in this way will be more biologically plausible and avoid being a weak analogy of the process on which they are based.

In (Andrews and Timmis 2005; Andrews and Timmis 2007), we explored how AIS have been developed in recent years focusing on their immunological inspirations. This led us to suggest that actively seeking out new immune theories for providing AIS inspiration could be of benefit. One such theory was identified as Cohen's (Cohen 2000) cognitive immune system, which was shown to incorporate many appealing properties that could inspire an engineering system, such as the *degeneracy* of antigen receptors and *patterns of response*. Antigen receptor degeneracy is the "capacity of any single antigen receptor to bind and respond to (recognize) many

P.S. Andrews (✉)
Department of Computer Science, University of York, Heslington, York, UK

D.R. Flower et al. (eds.), *Bioinformatics for Immunomics*, Immunomics Reviews 3,
DOI 10.1007/978-1-4419-0540-6_9,

different ligands" (Cohen et al. 2004). This, however, contradicts the classical view that immune specificity arises from the specific nature of antigen receptors. Despite this, Cohen (Cohen 2000) claims that specificity can arise from patterns of responding detectors such as T cells. For example, a pattern can emerge toward a particular antigen from the overlapping reactions of a population of degenerate immune receptors. Even though each immune receptor is nonspecific to its target, the result of all the receptor reactions together will be unique, and thus specific to that antigen.

Following the ideas of Stepney et al. (Stepney et al. 2005), the work presented here looks at how we can build an AIS based on the investigation of a suitable immune model. This model is inspired by the ideas of degeneracy and patterns of response, and looks at how the adaptable lymphocyte hypothesis of Grossman (Grossman and Paul 1992; Grossman 1993) can be used as a mechanism of producing a specific pattern of response from a population of degenerate detectors. In the next section, we outline the adaptable lymphocyte hypothesis, describing the tunable activation threshold (TAT) model. Based on this, we then investigate the behaviours of this model in a computational setting. This is followed by showing how we might combine the individual responses of degenerate tunable detectors to produce a single population pattern. From the insights gained by all these investigations, we detail a framework for incorporating the action of degenerate tunable detectors and population patterns into an AIS. This framework is then instantiated by developing an AIS for pattern classification.

The Adaptable Lymphocyte Hypothesis

In response to a number of observations that contradict the classical view of immunological tolerance, Grossman (Grossman and Paul 1992; Grossman 1993) presents the adaptable lymphocyte hypothesis. This states that the "responsiveness of individual lymphocytes to antigen and other signals can be tuned and updated" (Grossman 1993). The hypothesis is used by Grossman (Grossman and Paul 1992) to derive the TAT model, which we describe in this section. The basic assumption of the TAT model is that the result of an external stimulus on a cell is a change in its metabolic state that can be represented quantitatively. From this assumption, lymphocyte activation thresholds are proposed to be tuned internally by the cell based on the history of its environmental stimulation. The key definitions from the TAT model are:

- *Excitation*: A quantitatively expressed change in the metabolic state of a cell induced as a direct result of an external stimulus
- *Excitation Level*: A positive scalar measure of the excitation
- *Excitation Index*: A time-dependent, weighted average of the past excitation levels of the cell
- *Perturbation*: The difference between the current excitation level and the excitation index upon an excitation event
- *Activation Threshold*: The excitation index plus a fixed critical value

Grossman (Grossman and Paul 1992) suggests that the excitation index at time *t*, *I*(*t*), could be related to the excitation level, *E*(*t*) via the following equation:

$$\frac{\mathrm{d}I(t)}{\mathrm{d}t} = \alpha E(t)[E(t) - I(t)] \tag{1}$$

where α is a positive constant, and *E*(*t*)−*I*(*t*) represents the perturbation. This equation is derived from the idea that the excitation index could be measured using the concentrations of molecules undergoing storage-degradation cycles during excitation events. For example, the opposing action of tyrosine kinases and phosphates is suggested as a candidate. Grossman (Grossman and Paul 1992) states that if we suppose the production and degradation rates are both increasing functions of the excitation, this can be expressed as:

$$\frac{dC(t)}{\mathrm{d}t} = E(t)[\alpha E(t) - \alpha C(t)] \tag{2}$$

where *C*(*t*) is the molecule's concentration at time *t* and α and β are constants. If we denote $\alpha/\beta C(t) = I(t)$, then (2) reduces to (1) as so:

$$\begin{aligned}
\frac{dC(t)}{\mathrm{d}t} &= E(t)[\beta E(t) - \alpha C(t)] \\
\frac{d\alpha / \beta C(t)}{\mathrm{d}t} &= E(t)[E(t) - \alpha / \beta C(t)] \\
\frac{d\alpha / \beta C(t)}{\mathrm{d}t} &= \alpha E(t)[E(t) - \alpha / \beta C(t)] \\
\frac{dI(t)}{\mathrm{d}t} &= \alpha E(t)[E(t) - I(t)]
\end{aligned} \tag{3}$$

The excitation index, *I*(*t*), is therefore proportional to *C*(*t*).

An example of the dynamics of the TAT model is shown in Fig. 1, which has been reproduced from (Grossman and Paul 1992). This shows six distinct perturbations to the excitation of a cell over a period of time, with the corresponding evolution of the excitation index and activation threshold traced. Each perturbation is labelled along with a sign (+ or −) highlighting whether the excitation has breached the excitation index. For activation to occur, the excitation level must exceed the activation threshold. Of the six events shown, only the fourth would lead to activation. At time t_s, activation can no longer occur as the activation threshold has exceeded the saturation level, which is defined as the cell becoming temporarily *anergic*. Importantly, this definition of anergy is a reversible cellular process, which contradicts the classical definition of the term in immunology. Once the cell's excitation index drops sufficiently, it will become reactive again.

In addition to the TAT model, Grossman (Grossman 1993) incorporates tunable excitability into the adaptable lymphocyte hypothesis, defining it as "a measure of the cell's capacity to communicate with other relevant cells." It is described as being directly promoted by excitation events and an enhanced excitability facilitates

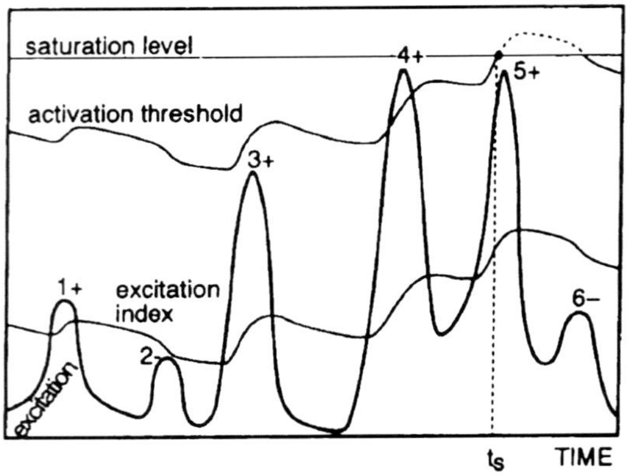

Fig. 1 Original tunable activation threshold behaviour from (Grossman and Paul 1992)

the ability of the cell to proliferate and differentiate. Unlike with the TAT model, no equation is given to describe the behaviour of excitability.

The adaptable lymphocyte hypothesis has a number of appealing properties from an engineering point of view. This has also been subsequently identified by Guzella et al. (Guzella et al. 2007) who suggest using tunable T cell thresholds as part of an AIS aimed at temporal anomaly detection. They highlight an initial architecture for a T cell inspired anomaly detection system, although the TAT model is not yet integrated into their architecture. They also highlight the work of a number of TAT models from theoretical immunology (Carneiro et al. 2005; van den Berg and Rand 2004; Scherer et al. 2004). Unlike these theoretical models we are not concerned with whether the TAT (1), or any other more complicated equation, is biologically accurate, but whether the adaptable lymphocyte ideas of Grossman (Grossman and Paul 1992; Grossman 1993) can provide suitable inspiration for AIS. In particular, can the adaptable lymphocyte ideas be a way of overcoming the property of degeneracy and providing specificity through a population response.

Investigating TAT Behaviours

The purpose of the investigations that follow in this section are twofold. Firstly, we analyse the TAT (1) to understand its behaviour, and whether this conforms to what has been stated by Grossman (Grossman and Paul 1992; Grossman 1993). Secondly, we investigate how it can be applied to data of the form typically used in AIS. As our goal is to translate the qualitative behaviour seen in the TAT model into an engineering domain, we have built a simple model and simulator that has been used for the experiments presented below. This model contains two main components: A population of *detectors* with tunable excitation indexes; and a population of *antigens* used to stimulate the detectors and provide the excitation. The model is iterative and at each time step, t, each detector is exposed to an antigenic stimulus and updated. The following equation (equivalent to (1)) is used to update the excitation index:

$$I(t+1) = I(t) + \alpha E(t)[E(t) - I(t)] \tag{4}$$

where the symbols are the same as those described for (1). An activation threshold, A_T, is also computed, which is the excitation index plus a fixed critical value, which we have denoted as θ:

$$A_T(t) = I(t) + \theta \tag{5}$$

The simulation of this model gives us the flexibility to *qualitatively* investigate various properties of the adaptable lymphocyte hypothesis.

The TAT Equation

Figure 2 demonstrates how the excitation index of a detector will tune given a constant excitation stimulus. It also shows how the activation threshold would trace above this excitation index. According to the TAT model, the detector would be activated

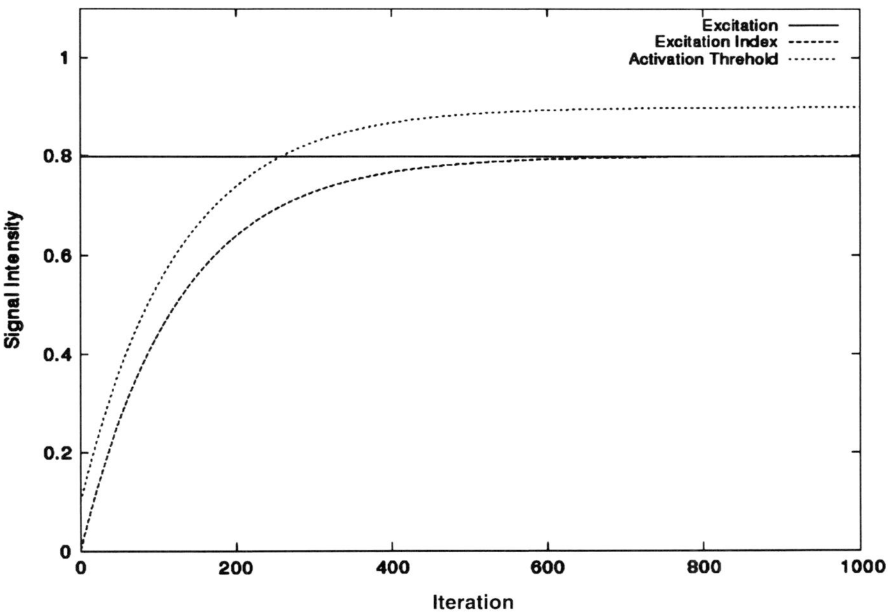

Fig. 2 An illustration of relationship between excitation index and an activation threshold set to trace 0.1 above the excitation index

between iterations 0 and around 250, where the activation threshold exceeds the value of the excitation. Figure 3 shows how the excitation index tunes to a variable stimulus that has been generated by a Gaussian distribution. This highlights the difference between the tuning and de-tuning behaviours of the TAT equation. We see that tuning occurs rapidly, whilst de-tuning occurs more slowly, especially as the excitation falls to a low value. These types of response behaviour are affected by the three main components of the update (4): A positive constant α, the size of the perturbation to the system $E(t)-I(t)$, and the excitation $E(t)$. In this section, we investigate each of these components, in turn, to assess its influence on the excitation index.

The α Parameter

The α parameter determines how quickly the excitation index tunes, the smaller it is, the smaller the increment to the excitation index and the slower it tunes to the value of the excitation. This parameter, therefore, controls the "memory effect" of the excitation index: The lower the α value the longer-term memory of past excitations. Figure 4 shows the effect of the α parameter has on the speed at which the excitation index tunes to a constant excitation. Three values of α are shown and as the value gets smaller, so the tuning gets slower. It is noted that the values chosen for α here are directly related to the number of iterations the simulation is run for.

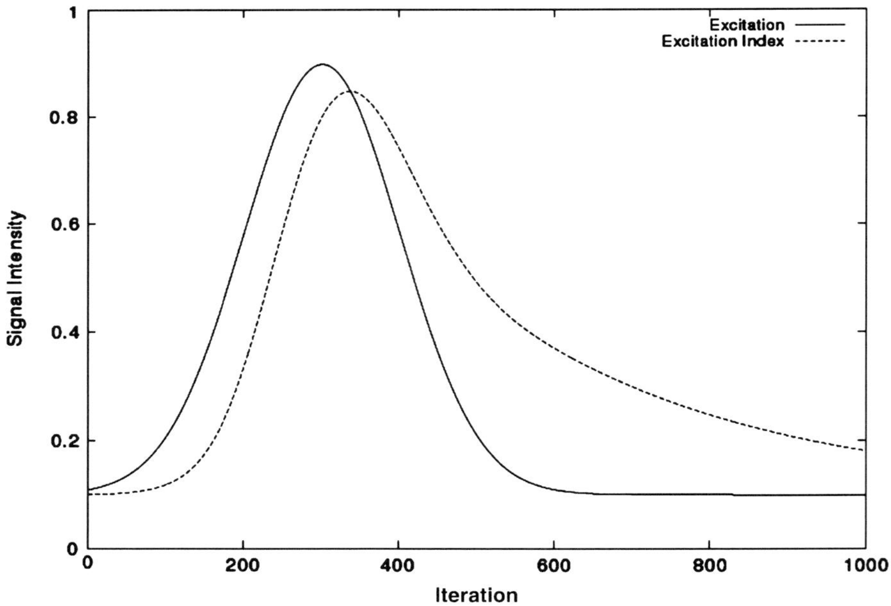

Fig. 3 An illustration of relationship between excitation and excitation index to a varying input stimulus

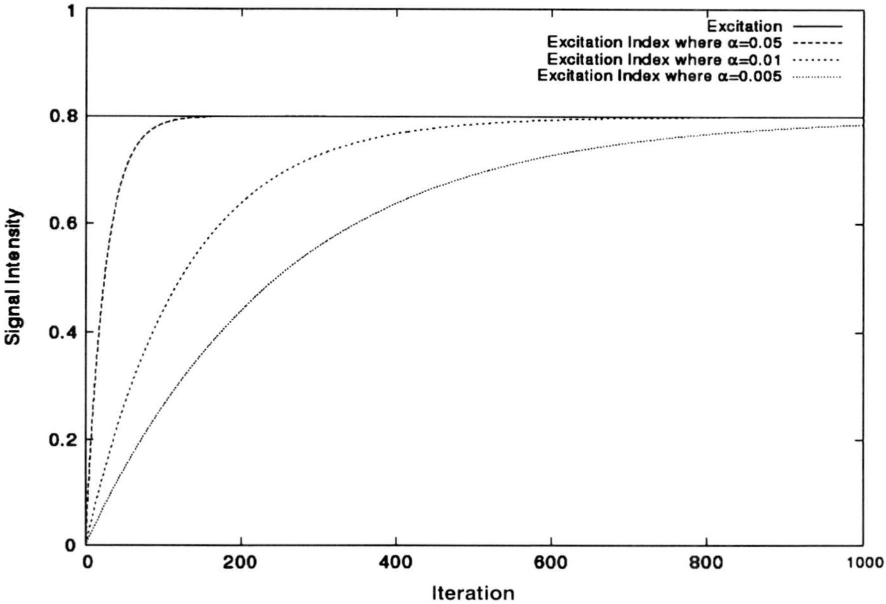

Fig. 4 The effect of parameter α on the speed at which the excitation index tunes to a constant excitation

The Perturbation *E(t)–I(t)*

This perturbation is the only term that can be negative, thus a positive perturbation will lead to an increase in the excitation index and a negative perturbation to a decrease. The larger the perturbation, the larger this term will be. We can see the effect of the perturbation in Fig. 3. At around 300 iterations, the excitation starts to fall. As it crosses the excitation index at around iteration 350, the perturbation becomes negative and the excitation index starts to de-tune.

The Excitation *E(t)*

Comparing Figs. 3 and 5 highlights the effect $E(t)$ has on scaling the excitation index depending on the level of excitation, which results in the excitation index tuning more quickly at higher excitations. The figures have exactly the same parameter for α and the same Gaussian shape to their excitations, but Fig. 3 has a baseline excitation of 0.1 to which the Gaussian is added, whereas Fig. 5 has a baseline of 0.05. Consequently the excitation index in the later figure tunes at a slightly lower rate. As the excitation value falls, so does the tuning of the excitation until at $E(t) = 0$ where no tuning occurs. In a biological setting, this lack of stimulus may be unlikely to occur; however, in an engineering context we need to be aware of this behaviour

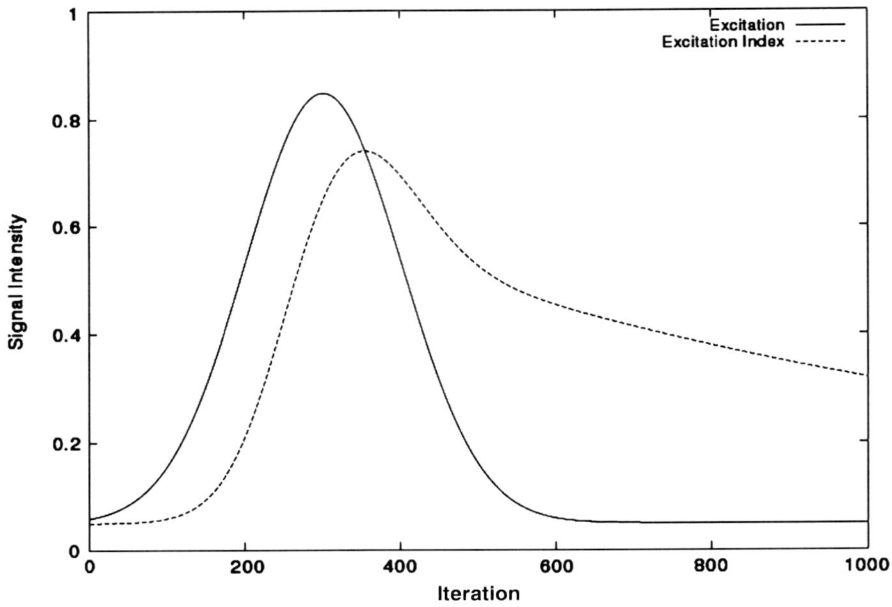

Fig. 5 Example of how the excitation effects the excitation index at lower excitation values. Compare with Fig. 3

as it could be an unwanted property if a constant background level of excitation can not be guaranteed. It is this term that makes the TAT (1) different from a simple sliding average of previous excitations.

Recreating Fig. 1

As a final example of the dynamics of the TAT equation, we generate an excitation distribution to visually match the original figure of Grossman (Grossman and Paul 1992) (Fig. 1) to assess if we can re-create the behaviours of the excitation index and activation threshold. This excitation distribution shown in Fig. 6 is created from the addition of six different- sized Gaussian distributions. The result provides a good visual match to Fig. 1. We conclude from these observations that the TAT (1) is suitable to produce qualitatively similar behaviours as those described Grossman (Grossman and Paul 1992; Grossman 1993) and summarised above.

AIS-like Data Example

Our next step in assessing the suitability of the TAT model for engineering, is to investigate its behaviour with a detector and antigen representation similar to those used in applied AIS. Typically this involves providing each with a string of symbols

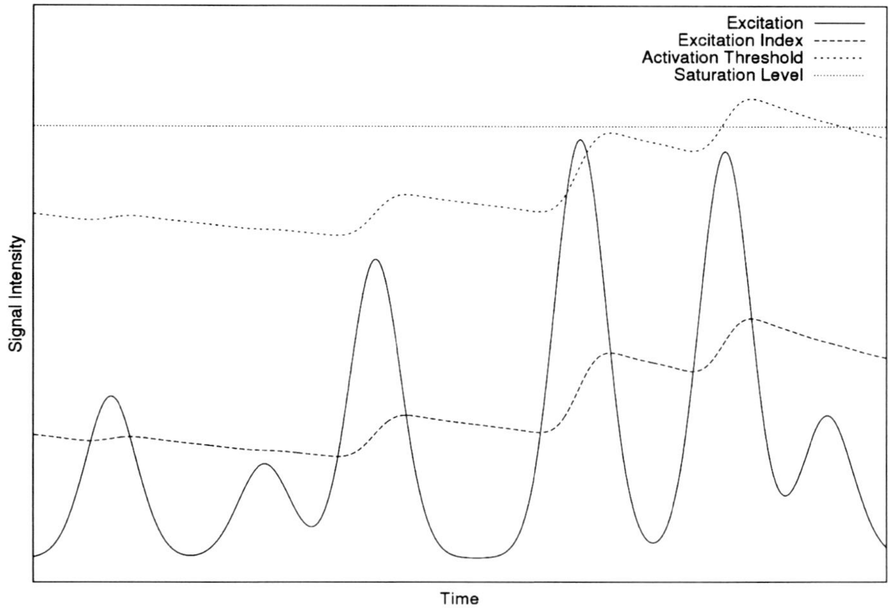

Fig. 6 Reproduction of TAT behaviours graph in Fig. 1 using (1), a combination of six Gaussian distributions for the excitation and $\alpha = 0.0075$

Table 1 Ten randomly generated antigen shapes

Antigen 0	[0.946848, 0.995084]
Antigen 1	[0.643525, 0.064986]
Antigen 2	[0.252777, 0.216995]
Antigen 3	[0.347241, 0.703525]
Antigen 4	[0.601634, 0.867903]
Antigen 5	[0.901560, 0.516287]
Antigen 6	[0.551903, 0.094621]
Antigen 7	[0.264536, 0.186120]
Antigen 8	[0.885042, 0.801183]
Antigen 9	[0.170462, 0.947705]

(normally, binary or real numbers) that represent the molecular shape of their binding regions. Here we investigate the simple case where antigen and T cell shapes are randomly generated real-valued vectors of size 2. For examples presented below, we generated 10 examples of each, which are presented in Tables 1 and 2, respectively.

Using the shapes shown in Table 2 and antigen number 0 from Table 1, Fig. 7 shows the excitation indices of ten tunable detectors tuning to a constant stimulus from the same antigen. At iteration 0 of the simulation,we set the excitation indices to equal 0. At each subsequent iteration the affinity between the antigen and detector

Table 2 Ten randomly generated T cell detector shapes

T cell 0	[0.843898, 0.511956]
T cell 1	[0.543662, 0.814270]
T cell 2	[0.002013, 0.963880]
T cell 3	[0.132122, 0.334890]
T cell 4	[0.676476, 0.204159]
T cell 5	[0.941392, 0.179290]
T cell 6	[0.780652, 0.046019]
T cell 7	[0.804552, 0.382894]
T cell 8	[0.659310, 0.571677]
T cell 9	[0.135789, 0.407200]

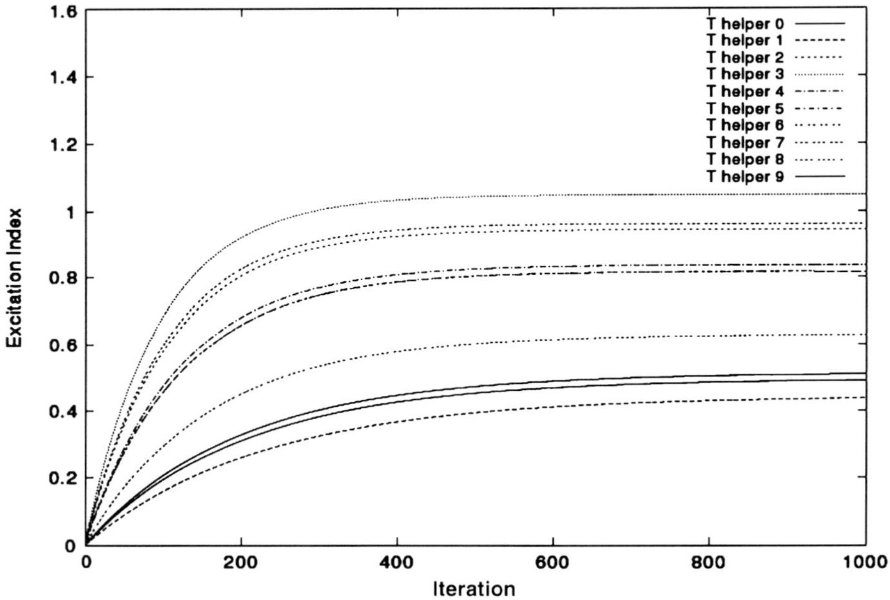

Fig. 7 The tuning of all T cell shapes from Table 2 to a constant stimulus by antigen 0 from Table 1. All detectors have $\alpha = 0.01$

shapes is calculated using their Euclidean distance, and is set to be the excitation for the detector. We can clearly see from Fig. 7 that each of the ten detectors tune to a different excitation index level, which is unsurprising as they are just tuning to their affinity for the antigen. We can also see that the excitation indices tune at different rates (a behaviour we explored above), with those detectors with lower affinity tuning more slowly. Results with applying different antigens from Table 1 to the same set of detectors show similar behaviours, with different detectors tuning to different levels at different rates.

Figure 8 shows how the same detector using the T cell shape number 1 from Table 2, responds to three different antigen shapes from Table 1. The antigen shapes chosen are antigen numbers 0, 2 and 7, as they represent the *most similar*

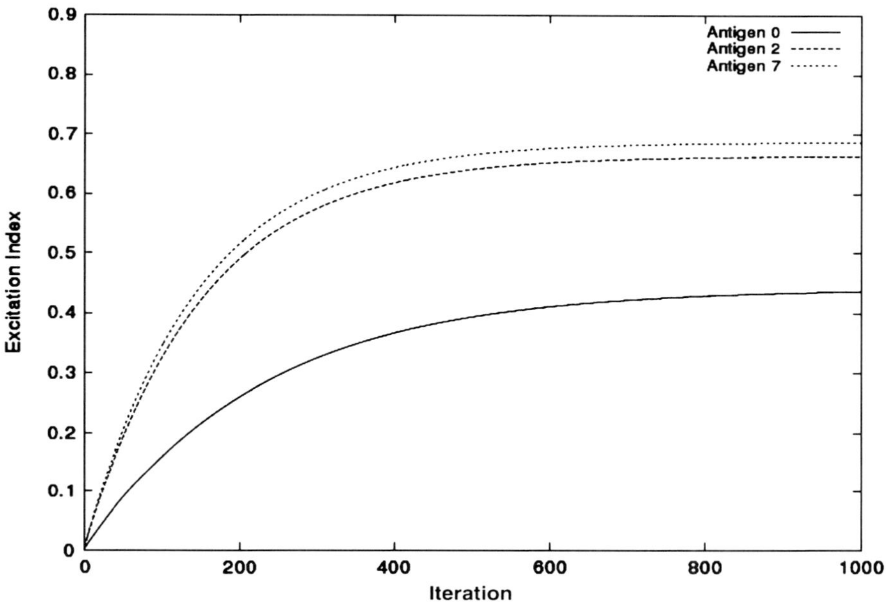

Fig. 8 The tuning of T cell shape 1 from Table 2 to a constant stimulus by antigens 0, 2 and 7 from Table 1 with $\alpha = 0.01$

(antigens 2 and 7) and the *most different* (antigens 0 and 7) pairings calculated by the Euclidean distance. From this figure we can clearly see that this detector produces a very similar excitation index response shape to the similar antigens but a different shape to the dissimilar ones.

These experiments with data of form used in AIS show a qualitatively similar excitation index tuning behaviour to the more abstract examples above. This re-enforces our view that the TAT (1) can be used effectively in an engineering setting.

Population Patterns

We have seen from the previous section how a single detector can tune its activation threshold according to (1). A population of these detectors can be applied to the same antigenic stimulus, and will each tune to a different value for their excitation index (and subsequent activation threshold). We also saw how a detector will tune to very similar levels to similar antigen, but to a different level to dissimilar antigens.

The question we now ask is 'how can we translate the reaction of the single detectors into a single pattern of response for the population of detectors?'. The way we have done this is to introduce the notion of a population size for each detector that represents the size of an entire *clonal population* of the detector. Our rationale is that in the real immune system, when lymphocytes are stimulated they start to proliferate.

Each detector is responding to the antigenic stimulus by a particular *degree* based on its affinity to that antigen as each detector is continually stimulated until it is driven into anergy and will effectively stop proliferating. As we have seen in Fig. 7 this will occur at different rates for each detector.

We can also make the speed at which detectors proliferate proportional to their current responsiveness. This idea comes from the idea of *excitability* described above. Here we saw that along with the excitation index, the excitability was responsible for a cell's ability to proliferate. Unlike with the TAT model, no equation is given to describe the behaviour of excitability, only that it is promoted by excitation events. Instead, just take the excitation index as a measure of the current responsiveness and multiply by a small constant to give a population increase at each iteration for an active detector.

Having this population expansion of a set of detectors gives two different dynamics to the response: The point at which the detectors are driven into anergy and stop being active; and the speed at which the detector population expands before this happens. Given a randomly generated population of these detectors we will get a clonal expansion of the detectors occurring at different rates, producing a pattern of population response that is typical of the antigen that induced it. Our population pattern generation algorithm is given by algorithm 1, where a typical stopping condition would either be until all detectors are no longer active or a pre-determined number of iterations.

Algorithm 1: Population response generation algorithm

```
input : i=data to be transformed
output : r=response pattern consisting of population dynamics over time
begin
    Randomly generated set of δ detectors, D
    repeat
        forall detectors, δ, in D do
            Calculate affinity with data item, i
            Set excitation level based on affinity
            Set excitation index using (4)
            if excitation>(excitation index+proliferation threshold, θ) then
                set detector active
            else
                detector not active
            end
            if detector is active then
                Increase population size proportional to excitation index, scaled by a
                    parameter μ
            end
        end
        Add active population value to response pattern r
    until stopping criteria has been met
End
```

To investigate the population patterns generated by our algorithm, we first look at Fig. 9. This shows the individual active population sizes for each of the 10 detectors with receptor shapes from Table 2 stimulated by antigen 7 from Table 1. We can clearly see the different responses generated with each antigen. The general trend here is for a detector with strong affinity (large excitation) to reach a large population size quickly, and then become anergic. This can be seen by the tall left-hand peak of detector number 2. Detectors with a weaker affinity will produce shorter peaks at later iterations such as detectors 3 and 9. It is interesting to note that even though detector 3 is driven to anergy after detector 9, it reaches a larger population size. This effect is controlled by the μ parameter that determines how the increase in size of the active population is scaled each iteration.

If we simply add the active populations of all detectors in Fig. 9 together, we then produce a single pattern of response for that set to an antigenic stimulus. This pattern of response simply shows a population size over a period of time. Figure 10 shows how such a pattern would look for the three antigens 0, 2 and 7 applied to the 10 detectors previously used. We see here how the population pattern of response is very similar for the two similar antigens (numbers 2 and 7 from Table 1) but quite different from the most dissimilar antigens (number 0).

There are four main parameters that affect the dynamics of the population pattern that is produced by algorithm 1. These are: The α parameter of the TAT (1); the number of detectors, δ; the proliferation threshold, θ; ; and the scaling factor for increasing the population size at each iteration, μ. Figure 11 shows the effect of

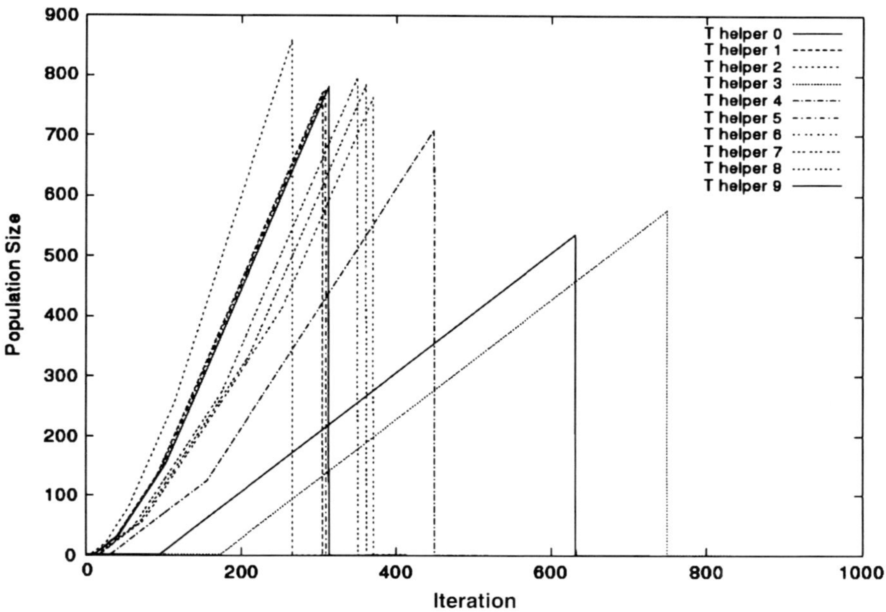

Fig. 9 Active population sizes of a set of 10 detectors being stimulated by the same antigen

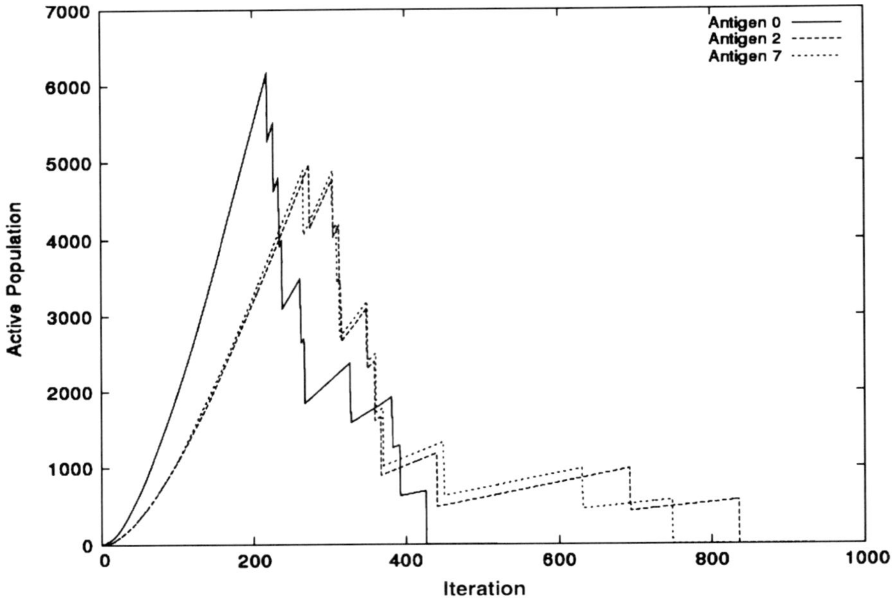

Fig. 10 The population response of three different antigens to the same degenerate tunable detector

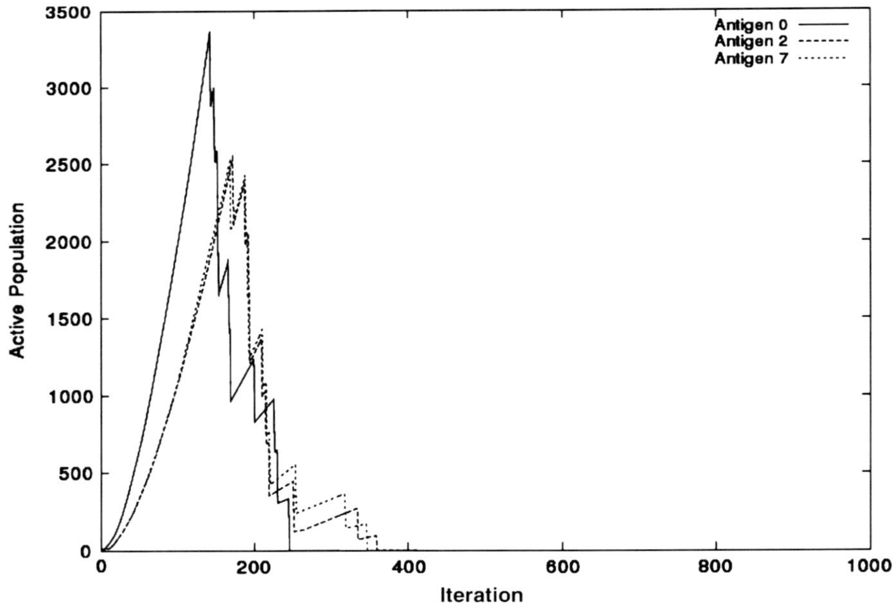

Fig. 11 The effect of increasing the proliferation threshold on the population response of three different antigens to the same degenerate tunable detector

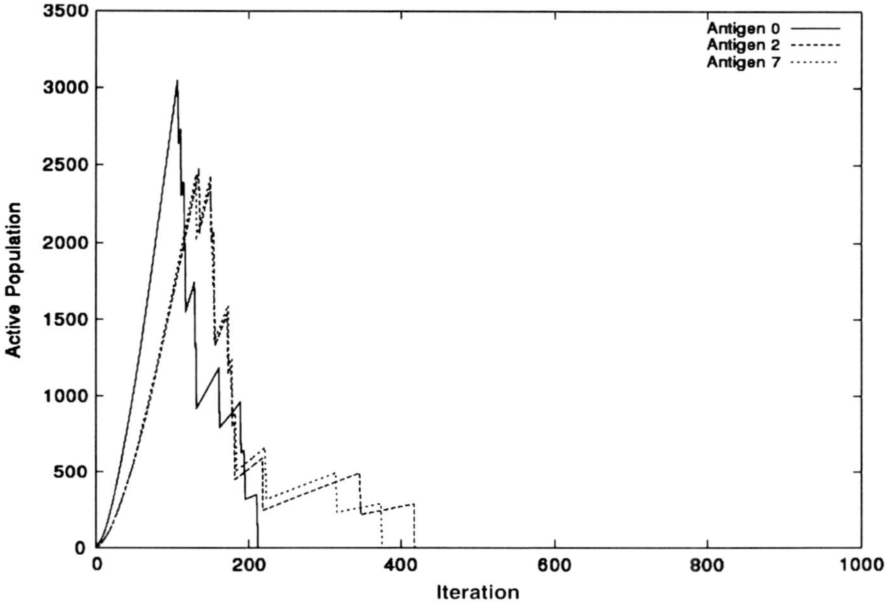

Fig. 12 The effect of increasing the α parameter on the population response of three different antigens to the same degenerate tunable detector

increasing θ on the same antigens and detector as Fig. 10. This details the effect of squashing the response shapes into a shorter number of iterations, although it is still possible to see distinct differences for the dissimilar antigens. Figure 12 shows the effect of increasing α on the same antigens and detector as Fig. 10. This highlights a very similar effect to the effect of θ. From this we can assume that by fixing θ to a small positive value (need simply to activate detectors), we can vary the shape of the graph by using just the α parameter.

The population response patterns seem quite insensitive to μ once it is above a small value. This is due to the parameter being an integer value, so when it scales the real-valued excitation, rounding off occurs. At low values of μ , this can have a noticeable effect. Finally, the δ parameter will simply produce graphs with the number of peaks equal to this value δ. The location of these peaks is determined by the randomly generated shape of the detectors and the effect of the other parameters just described.

A Framework for Degenerate Tunable Detectors

Having investigated the behaviour of the TAT model as above and identified a way in which we can produce a single pattern of response from a set of detectors, we have extracted a *framework* for the inclusion of degenerate tunable detectors into AIS.

We refer to this framework as the *patterns of degenerate tunable detector framework for AIS* (DTD-AIS). In this section, we first assess how the work presented above can be applied in an engineering context. Next, we present the DTD-AIS and lastly, we examine the parameters for the population pattern algorithm that is associated with our framework.

Patterns of Response for Engineering

It is clear that the generation of a pattern of response from an input data vector performs a *transformation* on that data from one representation to another. The output representation is affected by the algorithm and parameters that have generated it. Algorithm 1 above takes an input vector of any size and outputs a graph of population size over time. This population size graph will contain a number of randomly generated detectors that have been used to generate it.

Given that the generation of a populated response is performing a data transformation, it seems suitable for use as a data *pre-processing* mechanism. The data that has been processed can then be fed into another algorithm that performs some kind of application specific task on it. It is important to note that the data pre-processing stage by an algorithm such as algorithm 1 is itself independent of an application;however, the parameters you choose to generate the response may be tailored to the application.

This kind of pre-processing task has been described by Secker (Secker and Freitas 2007) as an instance construction task that is typical in data mining. They also point out that the immune inspired classification algorithm AIRS (Watkins et al. 2004) performs this task, pre-processing input data to be classified by a *k*-nearest neighbour algorithm.

The Algorithm Framework

The framework we have developed acts as scaffolding for incorporating patterns of degenerate tunable detectors as a data pre-processing mechanism is presented in Fig. 13. This contains seven main elements on which we will elaborate.

We start with an *application* that fulfils some sort of engineering requirement. This application will consist of *data* and an *algorithm*. The data will not be directly used by the application algorithm, but will be pre-processed. The output of this will eventually feed back as processed data to input into the algorithm. A population pattern generation algorithm (such as algorithm 1) will take as its input the application data and a number of specific parameters that will affect its dynamics. Based on knowledge from the application domain, these parameters can be chosen to best suit that domain, hence the application will influence parameter settings for the population pattern algorithm. The choice of population pattern algorithm needs to take into

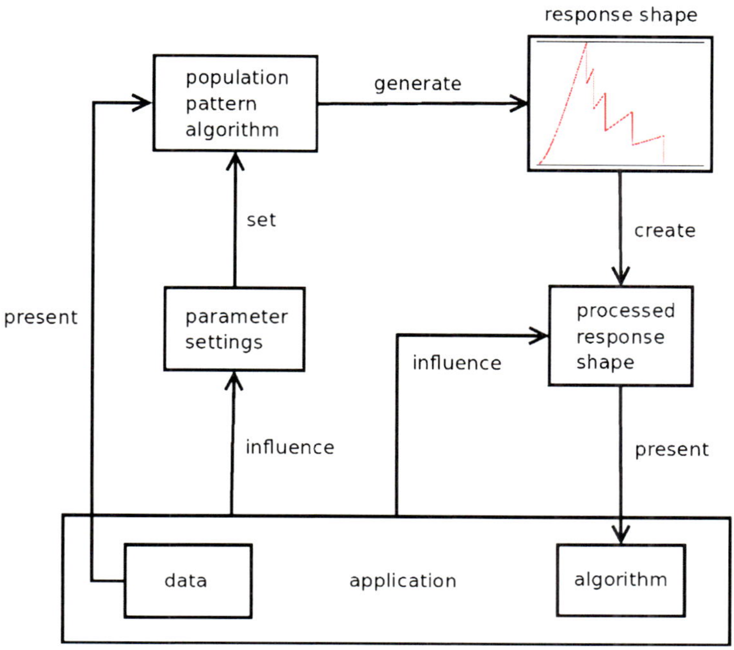

Fig. 13 An algorithm framework for incorporating patterns of degenerate tunable detectors in AIS

account engineering concerns such as the TAT model's inability to tune if the excitation is 0 (see above). This algorithm will take the input application data and generate a pattern response shape such as those seen in Fig. 10. This shape consists of an active population value (on the y-axis) plotted against a period of time (on the x-axis). It can then be processed to extract the interesting features. For example, we may not be interested in the whole shape but rather the elements of it such as: The height of the peaks; the iteration at which the peaks occurs; or the distance between the peaks.

Other processing of the response shape can also occur here, such as normalisation, or the combination of similar response shapes to produce an "average" response shape. Like the setting of population pattern algorithm parameters, the way in which the response shape is to be processed will be influenced by the application domain of the data. The output from the processed response shape will be presented as input to the application algorithm.

Parameter Settings for Population Pattern Algorithm

To instantiate our framework, we use algorithm 1 presented above as the population pattern algorithm. It is noted, however, that any suitable algorithm could be used to generate this pattern. In this section, we can outline possible parameter settings for this algorithm and how they will relate to the performance of the algorithm. When we talk about algorithm performance, we are considering two aspects: The application performance, which is the ability to get the "right" answer; and the run-time performance that affects much of the computational resources the algorithm will consume.

The α Parameter

This parameter affects the spread of the response graph along the time axis. A low level will give you a larger spread requiring more iterations, and therefore, increasing run-time performance. Too high a value will result in a graph where points are too close on the time axis adversely affecting application performance. A trade-off in the choice of α will probably exist between increased application performance and reduced run-time performance. From the investigations above, a suitable value would normally be in the range $0.005 \leq \alpha < 0.1$.

The θ Parameter

This parameter essentially gives a very similar control to the population behaviours as the α parameter. Therefore, we can fix this to a small positive value, which is required for detector activation, and then use α to control the shape of the population response. A suitable value is $\theta = 0.01$.

The μ Parameter

The μ parameter does not affect the shape of the population response if it is over a small value, other than to magnifying. Depending on how the response shape data is being used in the application, μ could affect the application performance. For example, with a high μ, the active population values could be on a far higher scale than those of the time. This should be taken into account. This is an integer parameter, and a suitable value is $\mu = 5$.

The δ Parameter

The more detectors we have, the more it will affect the run-time performance owing to the increase in function evaluations of the antigen. However, a greater number of detectors should have the capacity for more information to be encoded in the response. Like with the α parameter, a trade-off will probably exist between increased application performance and reduced run-time performance. A suitable value would be in the range $5 < \delta < 10$.

Instantiation of a Degenerate Tunable Detector AIS

In this section, we take our framework described in Fig. 13 and instantiate an AIS for a specific application. We first identify pattern classification as our application along with the relevant data to be pre-processed. Based on this data, we then identify the population pattern algorithm, its parameters, and how we will process the response shape. We then discuss our application algorithm, the k-nearest neighbour classifier. Our integrated classification AIS is then described, and we show some experiments and results.

Application and Data

It was mentioned earlier that the AIRS classifier performs a pre-processing of data before classification by k-nearest neighbour. We also note that the data used in pattern classification are typically of a form (data vectors) similar to the data representation we have used in the investigations above. We therefore identify pattern classification as a possible application area for instantiating our framework.

To investigate this claim further we used our simulator to produce population patterns of some real pattern classification data to visually inspect where we can generate similar response patterns for data instances of the same class. The data we chose for this are the classic pattern classification data set, Fisher's iris data set that contains 150 instances of three classes of iris: Setosa, virginica and versicolor. The data instances are vectors of length four. We applied five instances of each data class to the same five randomly generated detectors and examined the resulting response graphs. These are shown in Figs. 14–16 for setosa, virginica and versicolor, respectively. We can clearly see from these that the five instances of each class produce similar-shaped responses, and that each class looks different. This is especially true of the setosa class, with the virginica and versicolor showing more similarities, although they still appear to be different. This process of visually inspecting our graphs improves our confidence that we can use patterns of response in this context.

Algorithm, Settings and Response Shape

The previous section showed how the pattern response generation algorithm 1 was able to produce patterns that showed promise. We therefore use this algorithm as our pattern generation algorithm. The settings for this algorithm were explored in the context of engineering, so we also have an understanding of how to apply these. Additionally, we know the settings that were used to generate Figs. 14–16 so we can use these as a starting point for our experiments.

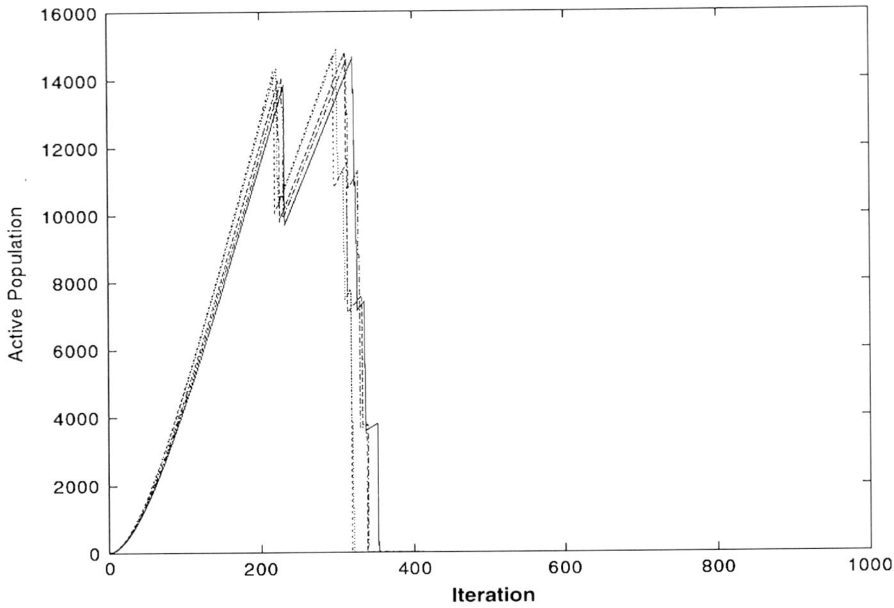

Fig. 14 Response shape of five instances of the setosa iris data class

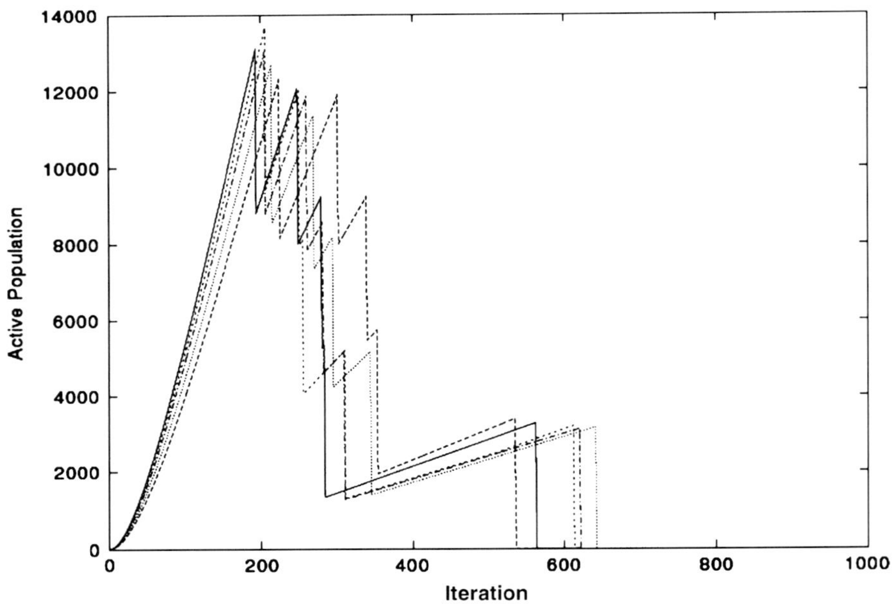

Fig. 15 Response shape of five instances of the virginica iris data class

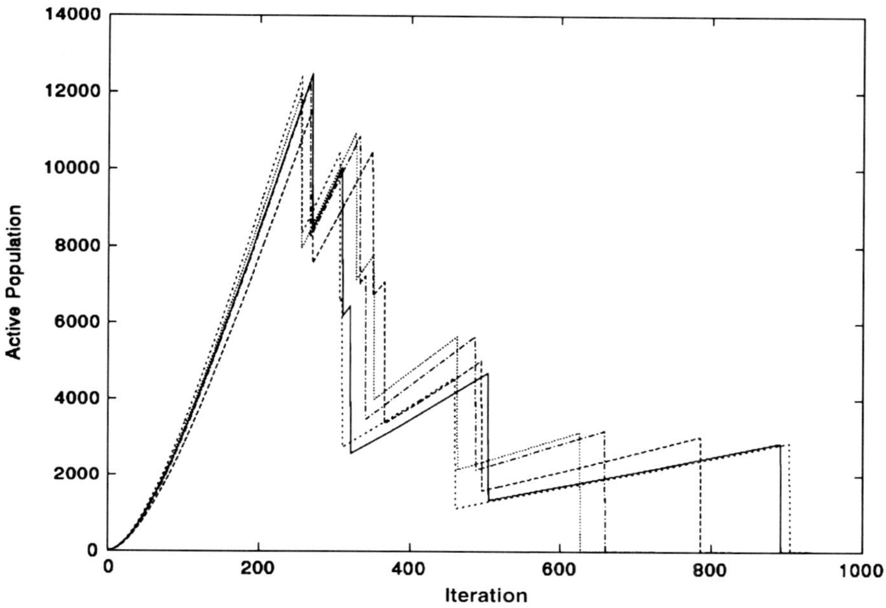

Fig. 16 Response shape of five instances of the versicolor iris data class

In order to use the response shape as input to the pattern classification algorithm, we process it first based on Figs. 14–16, which reveal data instances of the same class having peaks in similar areas. We are therefore interested in the relationship between these peaks. To produce a real-valued vector to feed into our application algorithm, we simply take the Euclidean distance between the peaks to produce a vector of length $\delta-1$ where δ is the number of detectors used to generate the responses.

k-Nearest Neighbour

Like AIRS (Watkins et al. 2004) we have chosen the *k*-nearest neighbour classifier as our application algorithm in the framework, which performs the actual task of classifying the vectors of response shape. *k*-nearest neighbour is the most basic example of an instance-based learning algorithm (Mitchell 1997) . It assumes that data instances map to points in an *n*-dimensional real-number space. The nearest neighbours of an instance to a set of other instances, is then defined by the Euclidean distance measure. If $k = 1$ then an instance is classified into the same class of its nearest neighbour. For values greater than $k = 1$, an instance is assigned to the class of the majority of the *k*-nearest neighbours. Consequently, an odd value of *k* is typically chosen to minimise the risk of an equal number of nearest neighbours belonging to different classes.

Freitas (Freitas and Timmis 2007) states that any algorithm performing generalisation must have *inductive bias*. This bias is simply a basis for favouring one hypothesis or data model over another, without which a choice could not be made. Mitchell (Mitchell 1997) points out that the inductive bias of a *k*-nearest neighbour algorithm corresponds to the assumption that the classification of an instance will be most similar to the classification of other instances that are nearby in Euclidean distance. From the visual examination of the patterns of response shown in Figs. 14–16, we believe this to be an acceptable assumption for our case.

An AIS Pattern Classifier

Having identified all the components of our AIS, we present the pseudocode in algorithm 2 for *patterns of response with tunable detectors AIS* (PoRTuDe). We input into our algorithm two sets of data, a testing set and a training set. The testing set comprises 20% of our total data set, and is used to calculate the overall accuracy of the algorithm. The training set is itself split into five equal- sized subsets so that we can perform a fivefold cross-validation during the training phase of the algorithm. This involves training the algorithm five times, each time leaving out one of the five subsets from the training data. Each training item is input to algorithm 1 and a response shape is generated and processed into a data vector of Euclidean distances between peaks. These vectors are added to the training response shapes. After all five training runs, the one with the best classification accuracy is chosen and used to classify the test data against using the *k*-nearest neighbour algorithm. The classification accuracy is then output for the testing data set.

Algorithm 2: Pattern classification algorithm PoRTuDe

input : The training and validation data, D_{train}, the test data, D_{test}
output : Classification accuracy on the test data
begin
  # Training Phase #
  forall training items, i_{train}, in D_{train} **do**
    Generate response shape, r_{train}, using algorithm 1
    Add r_{train} and its data class to set of training response shapes, R_{train}
  end
  Process values in R_{train} to extract Euclidean distance between graph peaks
  # Testing Phase #
  forall testing items, i_{test}, in D_{test} **do**
    Generate response shape, r_{test}, using algorithm 1
    Process r_{test} as required
    Perform *k*-NN classification on r_{test} using R_{train}
    Check classification and record success
End

Experiments and Results

We have applied the AIS shown in algorithm 2 to three well-known classification data sets: The iris data set, the ionosphere data set and the breast cancer data set from the UCI Machine Learning Repository (http://archive.ics.uci.edu/ml/). Each data set was split randomly five times producing five training-test sets, with the best reported split being presented in the results below. As a baseline we have also performed a simple k-nearest neighbour classification for each using values 1, 3 and 5 for k, the results of which are shown in Table 3.

We perform a parameter sweep for the PoRTuDe algorithm, keeping $\theta = 0.01$ and varying μ, δ and α in turn. Based on our analysis above, each combination of $\mu = \{2, 4\}$, $\delta = \{2, 5, 10\}$, $\alpha = \{0.01, 0.02, 0.04\}$ is tested. For each set of parameter settings, 50 runs are made and the mean, standard deviation, best run and worst run are reported. There is no stochastic element in the running of the algorithm itself, but the generation of the detector set is random, so the results are self-evident as to how well the system classifies and how robust the random detector generation is. The best results for each k-nearest neighbour and data set are shown in Table 4.

The results from Table 4 show that on average the PoRTuDe algorithm perform worse than a simple k-nearest neighbour for each data set and value of k. However,

Table 3 Classification results for k-nearest neighbour on the iris, ionosphere and breast cancer data sets

Data set	k value	Best accuracy (%)
Iris	1	96.67
Iris	3	96.67
Iris	5	96.67
Ionosphere	1	88.73
Ionosphere	3	85.92
Ionosphere	5	84.51
Cancer	1	97.1
Cancer	3	97.83
Cancer	5	97.83

Table 4 Classification results for PoRTuDe on the iris, ionosphere and breast cancer data sets

Data set	k value	Parameters	Mean (stddev)	Worst	Best
Iris	1	δ=10, μ=2, α=0.01	85.82 (7.15)	66.67	100.0
Iris	3	δ=10, μ=2, α=0.02	85.49 (7.58)	63.33	100.0
Iris	5	δ=10, μ=2, α=0.01	85.20 (8.82)	56.67	96.67
Ionosphere	1	δ=10, μ=2, α=0.01	73.66 (4.32)	69.014	83.10
Ionosphere	3	δ=10, μ=2, α=0.02	73.20 (5.12)	63.38	87.32
Ionosphere	5	δ=5, μ=2, α=0.04	73.80 (5.75)	66.20	84.51
Cancer	1	δ=10, μ=2, α=0.02	89.01 (2.90)	84.78	94.20
Cancer	3	δ=5, μ=5, α=0.02	90.40 (4.48)	79.71	97.83
Cancer	5	δ=5, μ=5, α=0.02	90.33 (3.89)	83.33	97.83

the results also show that the best runs of the PoRTuDe algorithm beat the k-nearest neighbour for the iris data set with a k of 1 and 3, the ionosphere data set for $k=3$ and the cancer data set for $k=3$. This shows that the generation of a pattern of response is possible to produce good pattern classification results, although it is quite apparent that there is a large variability in the results. This is due to the random generation of the detectors, and hence the algorithm is very sensitive to this aspect. The other negative aspect to the PoRTuDe algorithm is the higher computational cost compared to a simple k-nearest neighbour.

Conclusions

We have shown here how we can investigate a model of an immune mechanism from which we can then build an AIS. In particular, we have looked at the adaptable lymphocyte hypothesis of Grossman (Grossman and Paul 1992; Grossman 1993) to produce patterns of response from populations of tunable degenerate detectors. Based on this, we have shown in general that it can be used to perform a data pre-processing step in an AIS. An example AIS was then detailed that performed the task of pattern classification on data that had been pre-processed by this population pattern mechanism. Although the average classification results for the algorithm were disappointing, the best instances of the algorithm were able to out-perform the k-nearest neighbour. We can therefore claim that at the very least, our transformation of application data vectors into patterns of response does not destroy the underlying structure in the data.

Importantly, the process of investigating the TAT immune model free of any engineering bias give us a general understanding of the dynamics of the tuning equations and the effects of their parameters. We were then easily able to translate the model's ideas and parameters to the engineering domain and a specific engineering application. Knowledge of the model also enabled us to identify pattern classification as a possible application. These insights are of the kind that Stepney et al. (Stepney et al. 2005) suggested were the benefits that could be gained from understanding a biological model before developing any algorithm.

Future work can be identified for a number of areas. Firstly, there are a number of avenues for further investigation of the TAT and population patterns models. For example, we can look at: Different affinity measures between antigen and detectors; different ways of generating the response patterns; possible tools for analysing the graph patterns to enable us to predict what an "optimum" number of detectors might be; and ideas for nonrandom generation of detectors. Secondly, we can further analyse the PoRTuDe algorithm we have developed. This should focus on a more exhaustive analysis of the PoRTuDe parameters, and importantly, whether the sensitivity to the random detector generation can be addressed. We also plan to investigate other possible applications such as those with dynamic environments, for example continuous learning and robotics.

References

Andrews PS, Timmis J (2005) Inspiration for the next generation of artificial immune systems. In: Jacob C, Pilat ML, Bentley PJ, Timmis J (eds) Artificial immune systems, 4th international conference, ICARIS 2005. Vol 3627 of Lecture Notes in Computer Science, Springer, New York, pp 126–138

Andrews P, Timmis J (2007) Alternative inspiration for artificial immune systems: Exploiting Cohen's cognitive immune model. In: Flower D, Timmis J (eds) *In Silico* immunology. Springer, New York, pp 119–137

Carneiro J, Paixao T, Milutinovic D, Sousa J, Leon K, Gardner R, Faro J (2005) Immunological self-tolerance: Lessons from mathematical modeling. J Comput Appl Math 184:77–100

Cohen IR (2000) Tending Adam's Garden: Evolving the cognitive immune self. Elsevier Academic Press, San Diego, CA

Cohen IR, Hershberg U, Solomon S (2004) Antigen-receptor degeneracy and immunological paradigms. Mol Immunol 40:993–996

de Castro LN, Timmis J (2002) Artificial immune systems: A new computational intelligence approach. Springer, New York

Freitas AA, Timmis J (2007) Revisiting the foundations of artificial immune systems for data mining. IEEE Trans Evolutionary Comput 11(4):521–540

Grossman Z (1993) Cellular tolerance as a dynamic state of the adaptable lymphocyte. Immunol Rev 133:45–73

Grossman Z, Paul WE (1992) Adaptive cellular interactions in the immune system: The tunable activation threshold and the significance of subthreshold responses. Proc Natl Acad Sci USA 89(21):10365–10369

Guzella TS, Mota-Santos TA, Caminhas WM (2007) Towards a novel immune inspired approach to temporal anomaly detection. In: de Castro LN, Von Zuben FJ, Knidel H (eds) Artificial immune systems, 6th international conference, ICARIS 2007. Vol 4628 of Lecture Notes in Computer Science, Springer, New York, pp 119–130

Mitchell TM (1997) Machine learning. McGraw-Hill

Scherer A, Noest A, DeBoer RJ (2004) Activation-threshold tuning in an affinity model for the T-cell repertoire. Proc Royal Soc B 271(1539):609–616

Secker A, Freitas AA (2007) WAIRS: Improving classification accuracy by weighting attributes in the airs classifier. In: Proceedings of the IEEE congress on evolutionary computation (CEC07), IEEE Press, pp 3759–3765

Stepney S, Smith RE, Timmis J, Tyrrell AM, Neil MJ, Hone ANW (2005) Conceptual frameworks for artificial immune systems. Int J Unconventional Comput 1(3):315–338

van den Berg HA, Rand DA (2004) Dynamics of T cell activation threshold tuning. J Theor Biol 228:397–416

Watkins A, Timmis J, Boggess L (2004) Artificial immune recognition system (AIRS): An immune-inspired supervised learning algorithm. Genet Program Evolvable Mach 5(1):291–317

Defining the Elusive Molecular Self

Matthew N. Davies and Darren R. Flower

Introduction

The word "self" has many meanings; some harmonious, some in conflict. Properly differentiating between the self, as an unreified context able to encompass, or at least encapsulate, both the physical and psychological manifestations of identity, and the perhaps more mundane recognition of internal self ("me") versus external non-self ("you") is fundamental to many things. Yet outside of the dream world of philosophy there is much in the material world that can illuminate the nature of self. Perhaps the nearest we meet to an identical external self is the twin (Joseph 2002; Lykken 2006). Consideration of the twin is helpful when one is beginning an attempt to unravel the nature of self. Two kinds of twin are recognized: dizygotic (DZ or "non-identical twins") and monozygotic (MZ, or so-called "identical twins"). Dizygotic twins are also known as fraternal or biovular twins. DZ twins occur when two eggs are fertilized independently by different sperm. Globally, about 125 million human births are twins and triplets (1.9% of the world population), of which only 10 million are monozygotic (0.2% of the total population and 8% of twins). There are three types of dizygotic twin and two types of monozygotic twin. At 40% of twin births, female–male twins are the commonest; then female–female or sororal dizygotic twins, and finally male–male dizygotic twins. From the current perspective, there is little to interest us in the dizygotic twin; they are little different to other siblings and express their natural disparity. Monozygotic twins are, on the other hand, of clear interest. MZ twins occur when a single egg is fertilized, which subsequently forms a zygote, but then divides into two embryos; resulting fetuses will share the same womb. A zygote which splits within a few days of fertilization may result in cells which develop separately with its own placenta or chorion and its own sac or amnion. Such twins are dichorionic diamniotic (or di/di) twins; this occurs in about 1 in 4 cases. More usually, the zygote splits later, resulting in a shared placenta with two separate sacs. Such twins are known as monochorionic diamniotic (or mono/di) twins.

D.R. Flower (✉)
The Jenner Institute, University of Oxford, High Street Compton, Berkshire, RG20 7NN, UK
e-mail: darren.flower@jenner.ac.uk

D.R. Flower et al. (eds.), *Bioinformatics for Immunomics*, Immunomics Reviews 3,
DOI 10.1007/978-1-4419-0540-6_10,

More rarely, in about 1 in 100 cases, the zygote divides late enough to give rise to a shared placenta and a shared sac; this is called monochorionic monoamniotic (or mono/mono) twins.

When, in very rare cases indeed, a zygote divides extremely late conjoined or "Siamese" twins result. The two types of monozygotic twin, which are always the same sex, are, in terms of prevalence, female twins and then male twins; male MZ twins are the least prevalent of the five variations amongst twins. On very rare occasions, MZ twins may have different phenotypes (for example due to the deactivation of different X chromosomes in MZ female twins) or, as a result of aneuploidy, twins may possess different sexual phenotypes. This is usually a result of an XXY Klinefelter's syndrome zygote splitting unevenly. A variant form of monozygotic twins is the mirror image twins, where twins develop reverse asymmetric features; about 1 in 4 identical twins are mirror image twins.

Another, largely hypothetical, form of twin is the polar body twins, where an egg splits and the resulting cells are fertilized simultaneously by two different sperm. The resulting eggs develop into a chimeric blastomere. If this blastomere then undergoes twinning, two embryos will result. Each will have identical maternal genes but different paternal ones. Cells in each fetus carry genes from either sperm, resulting in chimeras. This results in so-called semi-identical or half-identical twins.

Barring mutations, monozygotic twins are assumed to have essentially identical genomes, although their appearances and other physical and mental characteristics can diverge significantly for environmental – accidents, etc. – and epigenetic reasons. Although monozygotic twins appear identical, at least superficially, people who know them well can usually tell them apart. MZ twins will not share the same fingerprints, which result from random, as well as genetic, factors; likewise, for retina prints. As they mature, MZ twins will tend to become less alike, as the number of epigenetic differences between MZ twins increases. Certain characteristics however, converge with age, such as IQ and personality. Monozygotic twins are of particular interest to geneticists seeking to determine differences between genetic and environmental factors in the development of human personality.

Let us return to our original central question. This question might be framed in several ways: do we know ourselves? How do we know ourselves? Do we see ourselves as others see us? No, yet it is just as true that we can effectively conceal from others our thoughts, our motives, and our intentions. Is it possible, therefore, that we can know ourselves? No, but neither can anyone else truly know us, as we know ourselves. There is an inevitable tension between one's inner identity and the image we present to others; the image is of itself complex, comprizing what we knowingly project, what propagates from that, and what we project but are not aware of. Yet, since we cannot fully conceal our self or see ourselves as others do, the totality of self is never fully grasped by an individual but only by some disembodied collective appreciation composed of the largely subjective and even arbitrary opinions of others. Self has several components: the physical manifestation of identity, but self, at least within a group or society, is also a collective property generated by others; yet self is not permanent and immutable – it changes and is highly dependent on context. As we shall see, definitions of immune self share many of these characteristics. In the present chapter, we shall attempt to explore the notion of molecular self; that diverse and synergistic combination of molecular patterns that comprize the language of immunological recognition.

Notions of Immunological Self

As we have explored above, self – the word and the concept embodied by the word – has many meanings and each meaning hosts many related explanations. We have also alluded to a definition of self that is pertinent to immunology, where the word relates to the ability of the immune system to identify molecules, cells, and organs as belonging to the host and to differentiate itself from non-self: molecules, cells, and organs of exogenous and potentially pathogenic origin. However, rather than say definition in the singular we should rather say definitions in the plural, for within immunology there are many definitions of self. In what follows we shall examine some of the multitude of ways of thinking about self and non-self.

However, before beginning we should perhaps first ask ourselves an important question. Why – apart from its immanent interest – is any of this important? Annually, 2 million people die of diseases that vaccines should prevent; 27 million children every year fail to receive basic vaccines (Flower 2008). Understanding self, and particularly the discrimination between self and non-self, lies at the heart of our attempts to understand the way vaccines work and how we are able to deceive the immune system into misinterpreting the signals it receives.

As we have seen, the idea of self, and thus non-self, is a pivotal metaphor prevalent in a wide variety of disciplines of human inquiry; the immune self is thus loaded with a tranche of associations derived from this diversity of fields of study. Fascinating parallels exist between the immune self and the myriad of definitions arising from the humanities, psychology, and molecular science. While we should not let these parallels distract us, we should nonetheless acknowledge them. For, by considering them, we highlight and make evident their importance, and further we help free ourselves from the confusing and misleading anthropomorphism so prevalent and pervasive within the conceptualization of science.

Much of the discussion of self is, of course, semantic. It revolves around arguments over the definition, and the use, of words. Like much else, this is deeply subjective, and such arguments are themselves riddled with prejudice, tendentiousness, and bias. In this context, one attempt to escape such constraints is provided by Polly Matzinger's danger model (Matzinger 2002). This proposes that the immune system reacts to danger signals, be they of external origin or from injured cells. Thus, the danger model effaces the immune self, replacing it with the idea of danger signals: any molecular signal of whatever origin that is itself dangerous or can act as a flag for the presence of other dangerous substances could act in this way. Such a model is simple and seemingly compeling. There are counter arguments, such as the role or inflammation, which is seen both as a cause and as an outcome of the mechanisms labelled as danger theory.

Clearly, both self and non-self can encode recognition signals. The self is thus encoded as being part of the host and non-self is encoded as being part of some identified non-host, a particular bacterium or set of bacteria for example. Alternatively, theories can be formulated wherein either but not both of the self or the non-self can be seen as empty placeholders. For example, a self could be identified as something possessing one or more signals of being part of the host; and non-self as being anything else. An entity is thus seen as "non-self" if it lacks a self-signal. The reverse could hold. Non-self could be identified as something possessing one or more signals of being part of a specific non-host organism; and self as being anything else. An entity is thus seen as "self" if it lacks a non-self-signal.

Thus, put simply or perhaps simplistically, we have three alternatives: a double-positive model, a self-positive model, or a non-self-positive model. There are many logical and practical problems with this third alternative. It would necessitate the existence of generic signals across all non-host organisms or a potentially infinite capacity for the storage of knowledge within the immune system. At least as far as we know, neither of these is, nor can be, realized in the context of a finite immune system. Obviously, a double-positive model allows for substances that are neither self nor non-self. The fecund minds of the reader will doubtless supply many other problems and objections implicit within these three alternatives. In reality, of course, things are not, nor ever would be, so simple. Reality is always more complex and less dignified than we hope. The real immune self is a composite, which, in part, exhibits features of all of these distinct alternatives.

A self-positive model is presented by Burnet's Clonal Selection Theory (Ada 2008). This posits that T cells which react to host epitopes are eliminated during development, allowing only T cells without reactivity to host epitopes to persist and thus engage the molecular products of alien organisms. Thus, the alien is extirpated by T cells and products they secrete, while constituents of the host pass unnoticed. The immune system sees the self as is an empty set and reacts only to non-self.

As with all theories, there are certain problems with that espoused by Burnet, many of these have been addressed theoretically and experimentally over the succeeding decades. In a sense, the "self" manifests itself as an implicit background against which alien substances – effectively non-self – can be identified: non-self stands out as distinct from this background and then thus be seen as different and needing to be eliminated. This theory displays certain similarities to ideas evinced by Ferdinand de Saussure (1857–1913) in his Course in General Linguistics (Ferdinand 1983). He states that in language, there are only differences, and that moreover although positive terms are required for a difference to exist, languages contain no positive terms only differences.

Parallels to immunology are obvious: self has within immunology no context-independent meaning. Self exists only as the complement of the non-self. Thus, one reading of Clonal Selection theory holds it to be purely differential and negative. However, difference and similarity are poorly formulated concepts in philosophy and this undermines many a theory, for similarity philosophers will often mean lack of identity. A more subtle and nuanced version of Burnet's theory will present it as deleted-self (a bounded set of molecular patterns) verses non-self (an open but incomplete repertoire of other – and distinct – molecular patterns).

The largely discredited network theory of the immune system owing to Jerne (Jerne 1971) is seen by many as a step beyond that taken by Burnet down the semiotic road laid by Saussure. Jerne introduced, or at least systematized, many now familiar terms: epitope, paratope, allotype, and idiotype. Jerne's contention was that the immune system was composed of a network or networks which were themselves composed of a large and complicated network of paratopes recognizing groups of idiotopes and networks of idiotopes recognized by groups of paratopes. An idiotope in this context is a set of individual epitopes. Jerne stresses the repression of lymphocytes; repression in the context of mutual recognition between immune agents. This suggests that immunity focuses inwards rather than outwards towards invading pathogens.

Non-self is seen as those objects that induce perturbations in the epitope–paratope network. There is no "non-self" and there is no "self"; there are only perturbations causing reorganization of the network. Non-self is any perturbation above a threshold. The designation of "self" and "non-self" is thus imposed by external observers, while the immune system sees only itself. Immunity is therefore not autonomous, and the immune self is defined by the regularity of interactions within the immune system. Non-self emerges from the internal or external disturbance of these regularities. Unfortunately, Jerne's theory suffers conceptual obscurity, particularly in how meaning is established. It also suffers from a lack of biological relevance. It is elegant but not supported by biology; a theory in search of supporting data. At a high level, it provides a retrospective explanation of events, yet is not predictive.

Irun Cohen argues that immune system can recognize the self as well as the non-self (Cohen 2007). There is no pivotal difference between self and non-self and that these qualities are contingent upon the interpretation made of them by the immune system. This interpretation is guided by genetic and somatic factors. Evolution has generated inflammatory mechanisms for coping with infection. This is innate immunity. Janeway clarifies and exemplifies this when he argues that immunity discriminates between non-infectious self and infectious non-self (Janeway and Medzhitov 2002). The somatic component is provided by the life history of the host organism. Mammals evolve more slowly than micro-organisms, and so pathogens would always win an immunological arms race. Cohen's arguments suggest that immunity is a highly orchestrated and contextual system which transcends the simple dichotomy between self and non-self.

Most theories are not explanations of the underlying biology but explanations of a sanitized theoretical abstraction: intellectually persuasive possibly, but certainly unable to explain the fuzzy and imprecise nature of nature itself. Most theories, and the abstractions they grow from, are built around certain key ideas which do indeed illuminate the issues at hand, yet it is only by addressing molecular mechanisms at the heart of immunology that we make this discourse of practical relevance and utility. Whether certain entities are viewed as "self" or as "non-self" emerges in context. The same agent is ignored in one context, yet may be attacked in another. Ivory tower discussions are only of interest to those who live in ivory towers. The wider world is interested in vaccines and drugs that save lives, reduce morbidity, and facilitate economic stability and growth.

Reductionist Approaches to the Immune Self

To reach a useful definition of the immune self, we need to journey back from our sojourn in the dream-world of philosophy, we need to re-engage with a materialist vision of reality, we need to deprecate the fantastical language of the philosopher and the semiotician, looking anew at the immune system as an emergent process arising ultimately from the interactions of molecules. This is clearly the realm of the reductionist, but the synthetic reductionist, who seeks to decompose, simplify, and explain the complexity of physical phenomena only to build it back into the complex and predictive models they construct. This is thus the realm of the biophysicist, the informatician, and the systems biologist.

In the view of the reductionist, which exemplifies the materialist core of the reductive approach to science, the biological basis of self must reside in the atoms and molecules of the biological systems under exploration. Indeed, the lowest level at which the notion – the very concept – of self retains meaning is the level of the molecule. Why is this? Simply put, because identity lies at the heart of the self, and, in a world that does not permit of homeopathic effects, identity exists at or above the level of atoms and molecules.

The "self" can, in the language of the late twentieth and early twenty-first centuries, be succinctly defined as a product or, more properly, a combination of genetic, epigenetic, and environmental factors working synergistically to create and define a whole. This whole begins but does not end with the genome. Outside of immunology, it is currently fashionable to think in terms of the genome as defining the self.

An organism's genome is the DNA sequence that encodes it. Part of a genome will code for genes; these genes will in turn make mRNA, which will make proteins. These proteins will undertake most, but not all, duties within the cell. Part of the genome regulates the expression of DNA as mRNA and proteins, rather than coding for them directly.

The current sequence of the human genome is a composite derived from five individuals. However, despite the enormous temporal and pecuniary investment in the science of genomics, even simple questions go unanswered. How many genes are there in human beings? This should be a simple question but it is not (Southan 2004). The putative size of the human genome has presently been revized down from figures in excess of 100,000 first to about 40,000 genes and then down to only 20,000. A recent and reliable estimate from 2006 puts the number of human protein coding genes at or about 25,043 (Nordström et al. 2006); while, a 2007 estimate places the value at about 20,488, with, say, another 100 genes yet to be found (Clamp et al. 2007). Each genome is different, and each genome can result in a diversity of phenotypes, each dependent, in part, on the environment.

In 2007, the first individual human genomes were sequenced and published (Levy et al. 2007). Thus James Watson and J Craig Venter became the first of thousands – perhaps in time millions – to know their own DNA. Such self-knowledge will, many hope, be a significant component driving the development of personalized medicine. Each unique individual is subtly different, and these differences result from the fore mentioned combination of genetic, environmental, and epigenetic factors. At the genetic level, most differences result from variant genes and from heritable diversity in gene regulatory mechanisms. Variant genes can include faulty genes that lead to genetic diseases and dominant and recessive alleles.

A lesson worth the learning is that somatic aspects of immunity are crucial for understanding many immunological behaviors and phenomena. We cannot easily reduce everything to genes received from our ancestors and antecedents via germline cells. Beyond genetic variation, there is epigenetic variation. Epigenetics we shall explore in a moment. Both genetics and epigentics are combined with effects from the environment – such as diet and environmental chemicals, including sex disruptors, drugs, and toxic pollutants – to elaborate and make manifest the fully autonomous systems from which biology is composed: cells, organs, and whole organisms. Calorific restriction and obesity can have dramatic effects on the survival of the individual organism; likewise for natural and artificial chemicals persisting in the environment.

Epigenetics is, to some extent at least, the nascent science of how genomes, through the medium of the organisms they code for, interact with their environment. It is a science and a phenomenon that has an immense and previously unrecognized impact upon the nature of self at the molecular level. In its modern incarnation, this idea – how a phenotype is contingent upon the interplay between environment, organism, and gene – was initially propounded by Conrad Waddington (1905–1975). It had been long assumed that such changes persisted for only the life of an organism, and were not inherited. The rational proposed was that gametogenesis would efface any such changes, with progeny inheriting a complement of unaltered, unaffected genes.

Several molecular mechanisms combine to effect epigenetics (Doerfler 2008). These include DNA methylation, so-called histone remodeling, and genomic imprinting. This occurs in mammals and higher plants, where there is a significant maternal investment in each offspring. In imprinting an allele from one parent is silenced. It happens for only a few genes. This mechanism probably evolved through competition over the allocation of resources to descendents, yet as only a single copy of each imprinted gene is inherited it is sensitive to any epigenetic modification induced by environmental change. Imprinted genes are typically involved in mediating metabolism and nutrient processing.

Epigenetic mechanisms allow changes in gene expression to respond to changes in the environment. Such change occurs both during the development of the embryo and throughout adult life. These changes allow cells to make, maintain, and transmit specific profiles of gene expression, even when cells divide. Such change is itself a vital requirement for the persistence and prosecution of complex and multi-cellular life. It doubtless pre-dates multi-cellular life, as it is also a feature of in prokaryotic and eukaryotic single-celled micro-organisms.

Epigenetic changes to gene expression are transmitted by so-called non-Mendelian mechanisms of inheritance. The evolutionary rationale for epigenetic inheritance suggests that it affords a rapid way of adapting to transient environmental changes without needing underlying genes to undergo Darwinian selection. Epigenetic mechanisms are themselves the product of genome evolution, creating alternative, additional mechanisms able to accelerate an organism's ability to adapt, survive, and, most importantly, to propagate itself by reproducing. Epigenetic alterations of gene expression may create novel phenotypes that, in turn, act through many generations, exerting selective pressure on some genes, stimulating long-term changes in the genome. Inheritance via epigenetic mechanisms may be viewed not as an evolutionary alternative complementing natural selection by mutation, but rather as a driving force behind durable and long-lasting genetic change.

Beyond the world of the genome – and even beyond the emerging epigenetic world – lies the world of the transcriptome, proteome, peptidome, glycome, and metabolome. Distinct proteins have different properties and thus different functions in different contexts. Identifying and elaborating the functions of innumerable gene-products using either high-throughput approaches or through traditional biochemistry, will be a more difficult and drawn out affair yet should prove rather more rewarding.

It was once thought that life could be understood by identifying each and every protein and then determining its function. Yet it has emerged that this was an over-simplified picture of cells and their behavior. Genes have many promoters and their expression is tightly regulated. Alternative splicing and RNA editing has replaced the one-to-one correspondence of gene to protein, and in its stead we see that one gene can be manifest as a plethora of alternate proteins.

A greater problem still arises from the burgeoning list of post-translational protein modifications (Sims and Reinberg 2008). Phosphorylation, for example, was discovered more than 40 years ago, and kinase and phosphatase cascades have subsequently been central to the decipherment of cellular mechanisms. Moreover, glycosylation, which labels proteins with both linear and branched carbohydrate chains, mediates a dauntingly complex modification system that regulates the ultimate function and cellular location of proteins within or beyond the cell, including the proteolytic half-life of circulating protein. Ubiquitination targets proteins for degradation. Reversible limited ubiquitination can modify the activity of certain proteins. Alternatively, ubiquitination can act as a timed suicide mechanism: the initial activation of the protein by the addition of the first ubiquitins is followed by a "switch-off" event as the proteasomal system recognizes the protein as a target for destruction. Ubiquitin can be modified at any of its seven lysine residues, generating different linear and branched chains leading to diverse functions. Other examples of PTMS include O-GlcNAcylation, poly(ADP-ribosyl)ation, and autophagy, a complex process requiring both protein–protein conjugation and lipidation. polyglutamylation, conjugation of the ubiquitin-like protein Nedd8, and the recently described Urm1, the most ancient protein-conjugation system.

Some of these modifications seem to be as abundant as protein phosphorylation. Moreover, they can conjugate to many sites on a single target and, in many cases, form chains of varying lengths. To make matters even more complex, they seem to frequently modulate each other. It is now clear that post-translational modifications (PTMs) generate enormous molecular diversity that allows for extensive fine-tuning of protein regulation and stability, making the task of understanding molecular processes a daunting one.

As our understanding of PTMs deepens, it will doubtless affect the labours of most, if not all, biologists. Scientists often seek to reduce problems to core concepts scrapped bare of other, less relevant information. This is can be useful when we are seeking the solution of short-term goals, but eventually, such artificial boundaries hinder our quest for knowledge. A systems-biology approach will be needed to integrate data into engaging hypotheses. Yet it is clear that this objective must be sought should we wish to comprehend life and its complexities. Such an understanding is, after all, the ultimate goal of research within bioscience.

Depending on your particular perspective, self is, ultimately, either a signal or something which is recognized. Likewise, ultimately, non-self is also simply just a signal or something recognized. That these signals are recognized and responded to within the context of a complex and sometimes confusing system, exhibiting confounding behavior on several levels, is irrelevant.

In some sense then, though how real this sense may be is open to question, the self is both molecules – peptides and proteins – and signals. The self and the non-self are merely, yet not solely, molecules that are recognized – or more properly bound – by other molecules. And that, as they say, is all that self and non-self ultimately are: molecules and their recognition by the host. All the rest is just waffle, confusion, and obfuscation. The signal in the self is the recognition event – MHC or antibody mediated – which triggers the immune system to respond.

It was once felt that the MHC would provide a simple, straightforward, unambiguous, and unequivocal criterion able to discriminate self from non-self. Class I MHC molecules are expressed by almost every nucleated cell and are able to act as the mediators of signals acting to identify self. This is realized in the ternary complex of peptide-MHC and TCR that is the necessary preliminary to the activation of the T cell and thus the initiation of concomitant immune responses.

Scientists with a philosophical bent often lambaste materialist reductionist approaches to the issue of self and non-self. They argue that only under certain conditions is there an unambiguous correspondence between a sign or signal and the thing being signified. They argue, and argue with some force and justification, that the meaning of a sign or the meaning of self is seldom exhausted by an inductively derived catalogue of instances of direct correspondence between signified and sign.

The self, and thus non-self, is dynamic. Under certain constraints and in response to certain stimuli, the immune system is capable of attacking host constituents. Autoimmunity is immunity turning against the self it is meant to defend. Autoimmunity is usually associated with disease, but can also be a normative function of homeostatic maintenance and control. The immune self is not a stable and unchanging entity but is clearly context dependent and prone to environmental influence.

The fundamental molecular mechanisms underlying cellular and humoral immunity are quite different. T-cell immunity is mediate by the molecular recognition of peptides bound to MHC molecules, essentially short denatured fragments excized from proteins via proteolytic degradation. B-cell mediated immunity is made manifest by antibody recognition of a protein antigen's 3-dimensional structure. Thus, the molecular recognition events at the heart of cellular immunity are essentially conformation independent and are instead mediated by recognition of amino acid sides within the context of a peptide–MHC complex. Humoral immunity is, by contrast, highly dependent on the conformation of a folded protein.

The Molecular Definition of Self: Innate Immunity

In a cellular context, the discrimination of "self" versus "non-self" by the immune system has largely focused on the recognition of fragments derived by proteolysis from host and pathogen proteins presented by classical MHC molecules. However, it has emerged that the innate system also plays an important nay a pivotal part in the sensing of non-self. The innate and adaptive immune systems are intimately connected and co-operate highly. The two halves of the overall immune system manifest a much greater and more sustained response to infection, by combining and integrating the optimal features of both systems. Protective immunity results from the interplay of the antigen-specific adaptive immune system with the more generic, less specific innate response. The recognition properties of the innate system do not exhibit optimization of specificity or selectivity. However, those of the adaptive immune system employ receptors that can undergo a refinement process that significantly enhances their capacity ability to recognize whole antigens or derived peptides.

Most of the operation of the innate immune system is preprogrammed, and uses widely distributed receptors able to recognize generic targets: conserved structural motifs or patterns which are characteristic of molecules common to pathogens and microbial life. It does this through the recognition, by "pattern recognition receptors" (PRRs), of so-called or pathogen associated molecular patterns (PAMPs) (de Diego et al. 2007). PRRs detect disturbances to the immune microenvironment (including discrimination of "non-self") and initiate appropriate innate responses (Areschoug and Gordon 2008).

It has been said that the crucial feature of PRRs is that they bind multiple ligands by recognizing common PAMPs rather than binding to unique if degenerate epitopes. Attractive though this may appear, it is not wholly accurate. The range of peptides bound by antibodies, and particularly by the MHC-TCR system, is much larger than is generally supposed. PRR engagement of lipopolysaccharide (LPS) or other PAMPs, elicits a response; this is typically pro-inflammatory, involving cytokine generation which activates immune cells. Such reactions are crucial to disease management, but must be controled since excessive responses are damaging.

Several distinct families of PRRs are known, (Areschoug and Gordon 2008; Kornbluth and Stone 2006). Arguably the most important, or at least the most prominent, are the so-called toll or toll-like receptors (TLRs). Humans have 10 TLRs; they sense both intracellular pathogens (viruses) and extracellular pathogens (bacteria and fungi). Some bind particular patterns contained in microbial DNA which are absent from vertebrate DNA. More specifically, ssRNA is recognized by TLR7 and TLR8 and dsRNA is recognized by TLR3. TLR2 and TLR6 recognize many ligands: bacterial lipoproteins, peptidoglycan, Zymosan, GPI anchors from *T.cruzi*, LPS and lipoarabinomannanm phosphatidylinositol dimannoside. TLR4 likewise recognizes LPS, Taxol, bacterial HSP60, F protein and fibronectin. TLR5 binds flagellin. TLR9, found on dendritic cells (DCs) and B cells, detects CpG motifs in DNA. An activated TLR-dependent signaling cascade ultimately induces expression of a variety of response molecules.

dsRNA is also recognized within the cytoplasm by another PRR – RNA helicases such as RIG-I. These are important PRRs, as are the cytosolic NOD-like receptors (NLRs), which play vital roles in innate immunity as intracellular sensors of pathogens and cell damage. This group includes NODs, NALPs, NAIP and IPAF. While TLRs signal from the cell surface or early endosome, NLRs are activated intracellularly by bacterial molecules, such as peptidoglycan, RNA, toxins and flagellin. Whole animal and cell-culture models of bacterial infection suggest a pro-inflammatory role for NLRs, including the regulation of cysteine proteases within the so-called inflammasome. The 700 kDa inflammasome is a multi-protein complex responsible for processing and secreting pro-inflammatory cytokines. Two types of inflammasome are known: NALP1 (comprizing NALP1, adaptor protein ASC, caspase 1 and 5) and NALP2/3 (comprizing Cardinal protein, ASC and caspase 1).

Other PRRs include FcγRs, which binds opsinized zymogen and serum amyloid P; CD35 and CD11b-CD18, which bind opsinized microbial cells; C-type lectins, such as the mannose receptor (which binds mannosyl and fucosyl moieties) and scavenger receptors (SRs). SRs bind leipoteichoic acid, degradation products from apoptotic cells, and Gram + ve bacteria. Other PRRs include MARCO, MER, PSr, CD36 and CD14.

PRRs are all able to recognize PAMPs, yet each receptor has unique binding properties, cellular expression and engages with different signaling pathways. This diversity within innate immunity protects us from a diverse spectrum of pathogens. PRRs are encoded by germ line genes. Since the structures of such receptors are inherited, resulting entirely from evolutionary pressure, their specificity is fixed. They evolve relatively slowly by the mechanisms of natural selection through standard processes of point mutation, gene duplication, and so on. The germ line nature of these receptors necessarily limits the eventual repertoire of recognition specificity exhibited by the innate immune system; it does not permit recognition of previously unknown antigens. Yet over long periods it can evolve to ignore self molecules and thus realize some robust discrimination between noninfectious self and infectious non-self.

Until recently adjuvant development was empirical (Aguilar and Rodriguez 2007); now new understanding of how adaptive immune responses are initiated has made rational development of a new generation of adjuvants possible (Bayry et al. 2008). The innate immune system plays a key role in controlling adaptive responses and thus activation of the innate immune system is essential for a strong adaptive response and the development of immune memory. This phenomenon is mediated by the interaction of evolutionarily conserved PAMPs with PRRs on cells of the innate and adaptive systems. PAMPs are usually shared by whole classes of organisms. LPS is, for example, found as a common component of gram negative bacteria. Some adjuvants act as immune potentiators, triggering an early innate immune response that enhances the vaccine effectiveness by increasing vaccine uptake. PAMPs can be seen as a subset of so-called "danger signals". Examples of danger signals include RNA or DNA, intracellular components released during necrotic cell death. Necrotic cells induce an inflammatory response which fosters and foments adaptive immune responses.

The innate immune system recognizes pathogens because they display evolutionarily conserved PAMPs, and purified or synthetic PAMPs exert potent adjuvant effects mediated by PRRs. PAMP-stimulated PRRs induce the maturation and migration of APCs, up-regulation of antigen-loaded class I and class II molecules, cell surface expression of co-stimulatory molecules and the production of cytokines and chemokines. Co-stimulatory molecules flag the microbial origin of presented antigen, help activate antigen-specific T cells and create an inflammatory environment that amplifies adaptive immune responses. Ligation of different PRRs can also modulate the type of immune response.

Importantly, there are also several small-molecule drug-like adjuvants, such as imiquimod, resiquimod and other imidazoquinolines. The discovery that PAMPs stimulate defined PRRs provides a strong impetus to the development of SMAs but many existing SMAs were discovered fortuitously (Kornbluth and Stone 2006; Bayry et al. 2008). For example, Levamisole, a DNA vaccine adjuvant, was developed as an antihelminthic, Bestatin, a tumour adjuvant, is an inhibitor of aminopeptidase N [CD13] and Bupivacaine, another DNA vaccine adjuvant, is a local anaesthetic. Imidazoquinolines (e.g., Imiquimod, Resiquimod, 852A, etc.), which target TLR-7 and-8, were developed as nucleoside analogues for antiviral or anti-tumour therapy. Other examples of nonmacromolecular adjuvants include monophosphoryl-lipid A, muramyl dipeptide, QS21, PLG, Seppic ISA-51 and CpG oligonucleotides. Optimized CpG oligonucleotides, which target TLR-9, are now entering late phase trials as adjuvants for the poorly immunogenic Hepatitis B vaccine. Likewise many proteins have been identified as potential adjuvant molecules.

The Molecular Definition of Self: Cellular Adaptive Immunology

Traditionally, when attempting to analyze and predict properties of the cellular response, immunoinformaticians have centered their attention and their efforts solely on the specificity of MHC molecules (Flower 2003; Flower and Doytchinova 2002). In more recent and more enlightened times, attention has turned to the richer, deeper, more challenging world of antigen presentation (Vivona et al. 2008; Davies and Flower 2007). However – and there is, it seems, always a however – a lack of lucent simplicity makes the subject appear confusing to the point of discombobulation and perplexing to the point of obfuscation. Our understanding of the manifold mechanisms underlying antigen presentation, and thus the manifestation of the molecular components of the immune self, is as yet incomplete and partial. These mechanisms, as we currently picture them, are by no means simple yet they certainly seem clever. As with all exciting science, many important aspects of these complex processes remain controversial.

The focus is and will remain the epitope, the immunological quantum that lies at the figurative heart of immunology, both as a science of description, and as a science of action. Immunology studies immune phenomena and the molecular mechanisms that underlie them in order to reach a full and proper understanding of these process in both practical and philosophical terms. As immune systems are not uncomplicated things, much laborious effort is still expended in enumerating and elaborating the copious detail implicit in such phenomena. Yet immunology, in the guise of vaccinology, is also a science of action, and much effort continues to be expended in trying to understand and manipulate non-self. Non-self, when manipulated properly, can yield effective and practically useful vaccines. The word vaccine can refer to all molecular or supramolecular agents able to stimulate specific, protective immunity against pathogenic microbes and the disease they cause. Vaccines act to mitigate the effects of subsequent infection as well as blocking the capability of pathogens to injure and kill their host organisms.

MHCs bind peptides, which are themselves derived through the proteolytic degradation of proteins. There are many alternative processing pathways, but the two best-understood are the classical Class I and classical Class II. MHCs are not indiscriminate binders, but importantly exhibit a finely tuned yet complex specificity for particular peptides sequences whose sequences are composed of the 20 commonly occurring amino acids. MHCs also display a wider specificity which is in itself quite catholic in terms of the molecules they can bind; MHCs are not restricted solely to peptides, they also bind a variety of other molecules.

The immunological peptide repertoire or immune peptidome is compounded in various ways, perhaps most significantly through the effect of PTMs. Such PTMs include phosphorylation, lipidation, and, most importantly, glycosylation. Glycosylated proteins can be targets for binding by cell surface receptors based on sugar binding leptin domains. Glycosylated epitopes can also be bound by TCRs and antibodies. Lipids can act as epitopes directly through their presentation by CD1. PTMs can also be transitory, such as phosphorylation, or more permanent, such as modified amino acids. Many of these can be part of functional epitopes recognized by the immune system. Glycosylation of a protein, for example, is dependent on the presence of sequence patterns or motifs (Ser/Thre-X-Asn or S/T-X-N for N-linked glycosylation and Ser or Thr for O-linked) but this is not enough to correctly predict them. If these motifs are present at solvent inaccessible regions of a protein rather on the surface then they will not be glycosylated. The specificity of glyscosylating enzymes are affected by residues surrounding these patterns; for example, Pro as the central residue in the S/T-X-N motif typically prevents glycosylation. Glycosylation is thus very context-dependent, and it is thus a system property of an organism, and can vary considerably in terms of the nature and extent of the different sugars that can become attached to proteins, at least in eukaryotic systems.

To these can be added a wide range of synthetically modified peptides, which are bound by MHCs and recognized by T cell, as well as natural and synthetic small molecules which can also bind MHCs. Small molecule drug-like compounds bind MHCs; this can mediate pathological effects and has important implications in behavior-modifying odor recognition.

Cell-surface Antigen presentation in the context of class I and class II MHC molecules demonstrates distinct differences. This arises in at least three different ways: one from the physical differences in peptide binding by the different classes; one from the TCR-mediated differences in the recognition of the two classes; and another from the significant differences in the complex machinery of antigen processing and transport that effects the conversion of whole proteins into fragmentary epitopes.

Class I MHC molecules make available to immune surveillance markers that sample important intracellular changes such as viral infection, the presence of intracellular bacteria, or malignant cellular transformation as seen in tumour cells. The flagging or signaling of such profound cellular events ensures the induction of an appropriate immune response by circulating CD8+ T cells. By contrast, class II MHC molecules reveal to circulating CD4+ T cell-mediated immune surveillance markers that sample extracellular events.

Cellular antigen presentation is affected significantly by the innate response. PAMPs and the PRRs that bind them affect both class I and class II antigen processing and presentation pathways. They regulate and orchestrate the spatio-temporal dynamics of MHC biosynthesis, antigen sequestration, and the reordering of the cytoskeleton.

The natural repertoire of class I MHC-presented peptides is rather broader than is widely supposed (Vyas et al. 2008; Lin et al. 2008; Loureiro and Ploegh 2006). MHC class I ligands are derived primarily from degraded endogenously expressed intracellular proteins. Intracellular peptide fragments arise from two sources: self-peptides derived from the host genome and proteins from external sources such as pathogenic microbes, principally those originating from viral infection. This seeming simplicity masks several layers of complexity.

There are several distinct steps in class I antigen processing and presentation. Antigens are initially acquired from proteins with errors, which may result from mis-incorporation or premature termination. Misfolded proteins are then targeted for degradation by being tagged with ubiquitin. The proteasome then proteolytically digests ubiquitylated proteins in a stochastic manner into a population of relatively short peptides. Subsequently, digested peptides are translocated from the cytoplasm into the endoplasmic reticulum by the transmembrane transporter-associated with antigen processing or TAP. Once in the ER, peptides are bound by newly formed class I MHC molecules. Class I MHC heavy chains and β_2-microglobulin are both synthesized in the ER. Formation of MHC–peptide complex is quite intricate and complicated, and it is facilitated by a variety of proteins including tapasin, calreticulin, and ERp57. Fully loaded trimeric complexes of class I heavy chains, β_2-microglobulin, and peptide permit optimal folding and glycosylation, and are transported via the Golgi complex to the cell surface.

MHC class I ligands are derived primarily from endogenous proteins and were previously though to be 8–11 amino acids long. However, there is now much evidence that long peptides (13–15 amino acids and above) are also presented by class I MHCs in many – possibly all – vertebrates including human, mouse, cattle, and horse. The repertoire of presented class I peptides is expanded through aberrant transcription and translation of viral and self-proteins. Various mechanisms pertain such as read through, which lead to alternative open reading frames, or alternative splicing, which generate protein isotypes with different sequences at exon–exon boundaries. Both of these have been shown to generate immunogenic epitopes. Autoantigen and self-tumor antigen transcripts for example experience elevated rates of alternative splicing. Alternative splicing is a major factor driving proteomic diversity and provides a partial rational for the gulf between small eukaryotic genomes and the enormous complexity of higher eukaryotes. Most alternative-splicing occurs in the coding region. The most commonly observed alternative splicing event is exon skipping in which an exon or set of continuous exons are permuted in different mRNAs. Less frequent are the use of donor and acceptor sites and intron retention.

Intracellular proteins, including newly synthesized proteins, are degraded quickly, producing large amounts of short peptides. Non-functional proteins, or defective ribosomal products (DRiP), result from errors in translation and processing. They form a significant proportion of newly synthesized proteins, which is rapidly digested by the proteasome. Viruses can invade host cells and generate viral proteins and bacteria can inject protein into the host cell via the type III secretion system; both are also degraded by the host.

Intracellular protein degradation is mediated by a multi-protein complex called the proteasome. A whole variety of protein including heat-denatured proteins, incorrectly assembled, mis-translated or mis-folded proteins, as well as regulatory proteins with limited half-lives, are targeted by the proteasome. For antigenic proteins, it favours oxidized protein substrates, since about 75% of oxidized intracellular proteins are degraded by proteasomes and the 20S proteasome prefers partially denatured oxidized proteins.

The proteasome is a multimeric proteinase comprizing a core of proteolytic enzymes flanked by a complex arrangement of regulatory elements able to recognize, amongst other things, a ubiquitin label. The proteosome has a site with trypsin-like activity (cleavage after basic residues), another with chymotrypsin-like activity (cleavage after hydrophobic residues), and yet another with peptidylglutamyl-peptide hydrolytic activity (cleavage after acidic residues).

There are several different forms of the proteasome found in the cell, including the three most crucial: the most basic form, the 20S core proteasome; the ATP-stimulated 26S proteasome; and the so-called immunoproteasome, which forms in response to certain viral infections. Vertebrate immunoproteasomes are assembled from three γ-interferon-inducible subunits that replace the constitutive subunits of the 20S version. The immunoproteasome has an altered hierarchy of proteosomal cleavage, enhancing cleavage after basic and hydrophobic residues and inhibiting cleavage after acidic residues. This is in accord with C-terminal amino acid preferences for class I MHC binding. Immunoproteosomes likewise leads to an increased production of octamer-to-decamer peptides that are believed to be optimal for binding to MHC class I molecules.

Misfolded, unfolded, or short-lived proteins are earmarked for destruction in eukaryotic cells by the attachment of small polymeric chains of ubiquitin, a small, highly conserved protein. The ubiquitin–proteasome system is an essential component of the cell's sophisticated quality control mechanisms necessary for proper maintenance of homeostasis. The ATP driven process of ubiquitination begins by forming a thiol–ester bond between the terminal carboxyl group of ubiquitin and the activated cysteine of the ubiquitin-activating enzyme, usually known as E1. Ubiquitin is then transferred to an ubiquitin-conjugating enzyme, called E2. Ubiquitin protein ligase, also referred to as E3, facilitates the transfer of ubiquitin to a lysine residue on the substrate protein. Additional ubiquitins are then added at lysine 48, again by E3, to form a polymeric chain; a length of four appears to be optimal. Ubiquitin protein ligases impose specificity on the process by bringing together substrate protein and the E2 enzyme. Polymeric ubiquitin chains can be linked not only via lysine 48 but also via other lysine residues. These variant ubiquitins divert protein substrates to fulfil diverse functions: DNA repair, mitochondrial inheritance, and also for cell uptake by endocytosis.

The mammalian 20S proteasome is a large, supra-molecular complex consisting of 14 copies of the α subunit and 14 copies of the β sub-units. This structure is responsible for degrading protein originating in both the cytosol and nucleus. The active sites of the proteasome's proteolytic enzymes are found on the β subunit on the inner face of the cylinder. The rings of α subunits form a barrier through which unfolded polypeptides enter; they also form the binding site of proteasome activator 28 (PA28).

The exact mechanism of proteasome cleavage is unclear, although enzyme-mediated protein cleavage seems to occur via the sliding of unfolded protein through the proteasome. The position of the peptide in the protein and the nature of adjacent sequences determines at some of the specificity of proteasome cleavage. In one experiment, a nonameric murine cytomegalovirus epitope was not cleaved when inserted into the hepatitis B virus protein, but was cleaved when a poly-alanine peptide was inserted next to it. There is much evidence to suggest that the proteosome is responsible for generating the C terminus but not the N terminus of the final presented peptide. The active sites of the proteasome have different specificities for the P1 residue of the peptide. The mammalian proteasome also cleaves after small neutral residues and after branched amino acids. Analysis of naturally cleaved peptides indicates that the residues on either side of the C-terminus and up to five residues flanking the N terminus can be related to proteasome cleavage.

After peptides are degraded by the proteasome, they are transferred into the lumen of the endoplasmic reticulum (ER). The translocation process from cytosol to ER consumes ATP. The so-called transporter associated with antigen processing, or TAP, is required for peptide transit. TAP also has the ability to interact with peptide-free class I HLA molecules in the ER. After peptides associate with class I HLA molecules, the resulting complexes are released from TAP and are then delivered to the cell surface.

TAP protein is a heterodimer and consists of TAP1 and TAP2. Both proteins are part of a family known as the ABC transporters. TAP is an example of a protein in the MHC assembly process which is encoded in the class II region. TAP1 and TAP2 associate in the ER and form the TAP heterodimer. The central region of the TAP protein is likely to be the binding site as polymorphic residues in rat TAP2 have been shown to contact the peptide and influence peptide selection and transport. The binding of peptides to TAP does not require ATP, while to transport peptide across the ER membrane does.

Immunology dogma suggests that newly synthesized class I MHC molecules are not stable in a peptide-free state and are retained in the ER in a partially folded form. Several chaperone proteins are needed to complete MHC folding. Newly synthesized MHC main chain is associated with a chaperone called calnexin, which is a transmembrane protein. It holds MHC molecules inside the ER. Calnexin may not be an absolute requirement for the assembly of MHC molecules as HLA molecules can be expressed in cell lines lacking calnexin. The immunoglobulin binding protein, or BiP, has a related function to that of calnexin, and may indeed replace it in the cell. When an MHC molecule has bound to β_2-microglobulin, it is released from calnexin and binds to two other proteins: tapasin and calreticulin (another chaperone which a function related to that of calnexin). After peptide binding, calreticulin and tapasin dissociate from the fully folded MHC molecule. Once complexed to peptide and β_2-microglobulin, the MHC protein leaves the ER and is transported to the cell surface. The peptide binding process is considered as the rate limiting step of MHC protein assembly as only a faction of the peptides are able to bind to MHC.

However, there are a number of other processing routes which complicate the simple picture outlined above. Peptides cleaved by the proteasome are 3–25 amino acids long, while most class I MHC epitopes are less than 15 amino acids long. Only about 15% of those peptides which are degraded by the proteasome are of the appropriate length for class I MHC binding. 70% of peptides are too short and 15% are too long. Long peptides may be trimmed to the correct size by various cellular peptidases. Analyses of peptide generation and T-cell epitopes expression in proteasome-inhibited cells suggest that cytoplasmic proteases other than proteasomes may also be involved in the antigen processing pathway.

Peptides are digested in the cytosol by several peptidases such as leucine aminopeptidase (LAP), tripeptidyl peptidase II (TPPII), thimet oligopeptidease (TOP), bleomycin hydrolase (BH) and puromycin-sensitive aminopeptidease (PSA). Tripeptidylpeptidase II (TPPII) was suggested to be a peptide supplier because of its ability to cleave peptides *In vitro* and its upregulation in cells surviving partial proteasome inhibition. Leucine aminopeptidase was found to generate antigenic peptides from N-terminally extended precursors. Puromycin sensitive animopeptidase and bleomycin hydrolase were shown to trim the N termini of synthetic peptides. Recently, an enzyme located in the lumen in ER and named ERAAP (ER aminopeptidase associated with antigen processing) or ERAP1, was shown to be responsible for the final trim of the N termini of peptides presented by MHC class I molecules. However, currently there is insufficient quantitative data about the role of these proteases to allow a precise bioinformatic evaluation of their impact on the antigen processing pathway. Alternatives to TAP are also now emerging, such as Sec61, which also effects retrograde transport back into the cytoplasm from the ER.

Class II MHC expression is believed to be restricted primarily to professional antigen-presenting cells (APCs), including macrophages and dendritic cells (DCs). In the MHC class II processing pathway, following the receptor-mediated endocytosis of exogenous antigens by APCs, presented proteins are targeted to the multi-compartment lysosomal–endosomal apparatus, passing first into endosomes, then into late endosomes, ending up in lysosomes. While in transit, antigens are proteolytically fragmented into peptides by cathepsins. Before final cell surface presentation, peptides are bound by class II MHCs. MHC class II ligands have a more variable length of 9–25 amino acids and are derived mainly from exogenous proteins. Peptide-bound class II MHC molecules are ultimately translocated to the cell surface where they are available for immune surveillance by CD4+ T cells.

Following ribosomal protein synthesis, the α- and β-chains comprizing MHC class II molecules associate with the so-called invariant chain (Ii or CD74). This helps ensure successful *In vivo* folding of the MHC, and thus protection of the peptide-binding groove. Within APCs, MHC class II heterodimers are co-synthesized with the so-called invariant chain (Ii): this obligate chaperone controls the intracellular trafficking of class II MHC molecules and helps regulate the binding of peptides by class II MHC. The repertoire of peptides presented by APCs by class molecules is contingent upon the stepwise proteolytic degradation of Ii to CLIP, the smallest fragment still able to bind class II-MHC. Another chaperone, HLA-DM (called H-2M in mice), catalyzes the exchange between CLIP complexed to MHC and degraded peptides. Class II MHCs then transit via vesicular transport to endolysosomal compartments, where they associate with endocytosed proteins.

External, extracellular antigen is endocytosed or phagocytosed by APCs and is directed into the phagosome. These membrane-bound organelles maturate, undergoing sequential modification, and ultimately fusing with lysosomes to create phagolysosomes, where ingested and degraded antigen encounter class II MHC molecules. MHC class II molecules that are contained in the lysosomes are loaded with peptide fragments formed by lysosomal proteases. Both MHC class II molecules and the tetraspanin member CD63 are specifically recruited to pathogen-containing phagosomes. Subsequently, MHC class II molecules, as well many other lysosomal proteins including tetraspanins, are transported in endolysosomal tubules to the cell surface. Surface MHC class II molecules can be found in membrane microdomains with other co-stimulatory proteins.

Peptide display is also dependent on the processing of proteins into peptides within the endosomal/lysosomal compartments. Here proteins, including those derived from pathogens, are degraded by cathepsins, a particular type of protease. Class II MHCs then bind these peptides and are subsequently transported to the cell surface. The peptide specificity of protein cleavage by Cathepsins has also been investigated and simple cleavage motifs are now known. However, more precise investigations are required before accurate predictive methods can be realized. Much still remains to be discovered in terms of the mechanism of class II presentation.

Immature DCs reside in tissues including the lungs, gastrointestinal tract, and skin. They play a vital role in activating naive T cells, undergoing a radical transformation when exposed to pathogens. DCs, and other professional antigen presenting cells, are largely responsible for presenting peptide fragments from antigens to immune surveillance by circulating T cells. The machinery of protein degradation, ultimately responsible for peptide presentation, is carefully controlled after DC activation. DCs exhibit several unique immune features, such as cross-presentation, where class I and the class II antigen presentation pathways converge, sampling the whole gamut of interacellular and extracellular antigens.

Whole pathogenic microbes exist transiently in interstitial space. Such pathogens are phagocytosed by DCs. In DCs, antigen presentation is tightly controlled by Toll-like-receptor (TLRs) signalling pathways. Cell Surface TLRs are activated by phagocytosis of bacteria. DC activation by TLR induces endolysosomal tubule formation and phagosome maturation. These subcellular structures contain a plethora of proteins, including class II MHC molecules, and effect protein delivery to the cell surface. Phagocytosis is restricted to professional APCs. It takes up microorganisms and apoptotic bodies and much else besides, and shuttles them into phagosomes. These undergo multifarious modifications, including protein recruitment and fusion with other vesicles. Fusion with lysosomes to create phagolysosomes is ultimate stage in phagosome maturation. In endocytic vesicles, viruses engage TLRs able to recognize nucleic acids. Presentation of antigen complexed with the TLR4 ligand LPS can activate T cells more effectively than antigen alone. Antigen degradation after lysosomal acidification of endosomal compartments is controlled rigorously in DCs. Acidification of this compartment allows optimal activity of lysosomal acid hydrolases and cathepsins.

Another complicating feature germane to this discussion is the issue of cross-presentation. This is again immunology jargon and really refers to the conceptually unalarming observation that the class I and class II pathways are not distinct but are actually interconnected or, at least, leaky, allowing peptides from one pathway access to the other. In the MHC class I pathway, peptides are generated by various flavours of proteosome and other major proteases, like TPII, in the cytosol. Recently identified amino-terminal peptidases in both the cytoplasm and the ER then trim these peptides. This process roughly parallels endosomal processing, in which endoproteolysis is followed by amino- and carboxy-terminal trimming. Many peptides can escape from the endosome and pass into the cytoplasm where they enter the familiar class I pathway comprizing the proteasome, TAP and MHC.

Likewise, endogenously expressed cytosolic and nuclear antigens access MHC class II via a number of intracellular autophagic pathways, including macroautophagy, microautophagy and chaperone-mediated autophagy (Menéndez-Benito and Neefjes 2007; Strawbridge and Blum 2007). Macroautophagy is characterized by the formation of autophagosomes, double-membrane structures, which capture proteins from the cytoplasm, processing them via acidic proteases, and eventually fuse with lysosomes. Microautophagy involves lysosomal invagination which directly sequesters cytoplasmic proteins. During receptor-mediated autophagy, proteases in the cytosol produce peptides that are translocated into lysosomes by a combination involving LAMP-2a, hsp, and hsc. These three processes all probably deliver cytoplasmic proteins and/or peptides to endosomal compartments for binding by MHC class II.

Taken together all these mechanisms greatly increase the potential size and diversity of the immunogenic repertoire – or Immunome – of reactive peptides. It is perhaps stating the obvious, but in the absence of infection, cell surface MHC class I molecules will, under steady-state conditions, only be associated with self-peptides. For example, the class I peptide repertoire of mouse thymocytes is largely derived from highly abundant mRNA transcripts; peptides from cyclins, cyclin-dependent kinases, and helicases were particularly abundant. About one in four class I peptides were different when taken from transformed thymocytes, with half originating in proteins involved in neoplastic transformation, such as those from the PI3K-AKT-mTOR pathway. Thus, one may argue, and argue cogently, that, in the face of such complexity, the only realistic way to address this potential enormity of the peptide repertoire is via computational analysis and prediction.

The Molecular Definition of Self: CD1 Presentation

The presentation of lipid-containing antigenic molecules by MHC-class-I-like CD1 has gone a long way to explaining how the immune system is able to recognize non-protein antigens (Florence et al. 2008; Mori and De Libero 2008; Zajonc and Kronenberg 2007; Barral and Brenner 2007). CD1 molecules are able to bind and present amphipathic lipid antigens for recognition by T-cells. The CD1 family includes various subtypes – CD1a, CD1b, CD1c, CD1d and CD1e – which present self and foreign lipid antigens. CD1 shares sequence and structural similarity to class I MHC molecules, and is also complexed with β_2-microglobulin (β_2m). CD1 is divided into three groups: group 1 contains CD1a, CD1b and CD1c; group 2 contains CD1d; and group 3 contains CD1e. CD1 molecules possess hydrophobic channels within their binding grooves allowing the aliphatic hydrocarbon chains of lipids to bind deep below their surface. In a process that shares similarities with MHC mediated peptide recognition, the exposed polar parts of bound lipids, together with surface regions of the CD1, are then bound by TCRs.

Group 1 CD1 presents a vast array of microbial lipids to clonally diverse T cells mediating adaptive cellular immunity. To survey the intracellular compartments of APCs for lipid antigens, CD1 molecules traffic from the ER to the cell surface and then back into the cell, where they circulate through various endocytic compartments before returning to the plasma membrane bearing lipid antigens. CD1 molecules first bind lipid antigens in the ER and then may exchange these for other self- or microbial-lipid antigens that are encountered along the endocytic pathway.

The ER loading may involve MTP, and ER-loaded lipids may be exchanged in lysosomes by saposins. Foreign lipid antigens may be delivered to APCs in phagocytosed microorganisms or in lipoprotein particles. Lipid–antigen–CD1 complexes expressed on the cell surface then activate both adaptive and innate-like lymphocyte responses. Group-1-CD1-restricted T cells appear to function as clonally diverse adaptive immune cells that expand following infection.

Group 2 (CD1d) presents lipids to natural killer T (NKT) cells. A subtype of NKT cells, the so-called invariant NKT (iNKT) cells, is CD1d-restricted and acts as part of the innate immune system. Such cells express an invariant T-cell receptor (TCR) α-chain. This receptor mediates a response within hours of activation following antigen recognition and is a potent innate immune activator. iNKT cell activation invokes broad immune responses by giving rise to the rapid release of cytokines and chemokines, which help regulate innate and adaptive immune responses to pathogens, as well as certain kinds of cancer and auto-immune antigens.

Recent studies of lipid antigens provide insight into this surprizing range of T-cell responses, which are closely linked to the distinct CD1 isoforms and the invariant and diverse T-cell responses they elicit. iNKT cells are activated by self antigens and by cytokines secreted by APC in the absence of pathogenic antigens: this is an entirely unlooked for mechanism of T-cell response without seeming parallel among MHC-restricted responses.

The Molecular Definition of Self: Humoral Adaptive Immunity

Another entity mediating recognition of self is the so-called B cell epitope. These are regions of the surface of a protein, or other biomacromolecule, recognized by soluble or membrane-bound antibody molecules. The protection offered by all vaccines except BCG is mediated completely or predominantly through the induction of antibodies, which act mostly in infection at the bacteremic or viremic stage.

B cell epitopes can be linear (also called continuous) or discontinuous. Linear epitopes are single, short, continuous subsequences within an antigen. Discontinuous epitopes are groups of individual, isolated residues forming patches on the surface of the antigen. The verity and exegesis of an epitope depends on the nature of their experimental determination. Linear epitopes are identified using some kind of experimental screening procedure, usually PEPSCAN, where by overlapping sequences are assayed against pre-existing *ex vivo* antibodies. Discontinuous epitopes are usually identified from the structure of an antigen, typically one derived experimentally by X-ray crystallography or multidimensional NMR. Discontinuous epitopes are also identified by making site-directed mutants of the antigen and testing them for their effect on antibody binding.

Sequence-based B cell epitope prediction methods are limited to the identification of linear epitopes. If we look back a decade or two, then most predictors of either T cells or B cell epitopes were based on identifying maximally valued regions of sequences – essentially looking for peaks, or troughs, in some form of a propensity plot. This was long ago shown to be inappropriate for T cell epitopes and consequently many advanced methods for T cell epitopes prediction have arisen. However, many – should that be most, if not actually all – B cell epitope prediction methods continue to rely, wholly or in part, on finding such peaks. However, no single property is known that is able to predict linear or discontinous epitope location with any reliability or accuracy. Most prediction methods use properties related to surface exposure – such as accessibility, hydrophilicity, flexibility/mobility and loop and turn structures – since it is believed that epitopes, at least for nondenatured proteins, must be solvent-accessible if antibody binding is to occur.

Early methods adopted a sliding-window method, adapting well used hydropathy scaled to identify maximal property peaks. A correctly predicted epitope will correspond to a peak within a few residues of an antigenic residue. Short window sizes were required to reduce erratic peak values; the optimal being six residues. Longer window sizes performed less well, perhaps because of an increasing probability of including hydrophobic residues. Using data-sets representing the most stringent examples of peer-reviewed publications describing linear epitope-mapped protein sequences, Blythe and Flower (Blythe and Flower 2005) have explored the validity of B cell epitope prediction using sequence profiles of amino acid scales. Using 484 amino acid scales and 50 epitope-mapped protein sequences, as defined using polyclonal antibodies, the analysis of both single sequence and combined profiles indicated in the most categorical terms that the underlying approach was of little or no practical value – in terms of ROC plot analysis, the best method produced predictions little better than random.

The poor performance demonstrated by BCE prediction algorithms is troubling. No explanation seems overly convincing. It is unlikely that available methodology is to blame, as data-mining techniques have proved much more successful in other areas. The explanation favoured here again targets the experimental data as the source of the problem. The most widely available data derives from PEPSCAN, and there are reasons to suspect that this not what it seems or people believe it to be. Experimentally derived epitopes are identified by assayed against pre-existing antibodies with affinity for whole antigens. However, when such "epitopes" are mapped back onto antigen structures their locations are scattered randomly through the protein. They do not form discrete patches as one would expect if they are simple mimics of crystallographically identified discontinuous epitopes. These *In situ* epitopes can be exposed or completely buried, and thus inaccessible to antibody binding, and also in every state in between. If we compare the conformation adopted by antibody bound peptides with those *In situ* in the intact antigen we see that they are typically very different. However, if we compare epitope peptides in intact antigen and in whole antigen–antibody complexes they are very similar. Thus the recognition of epitopes in a PEPSCAN analysis requires explanations other than the simplest one of a one-to-one correspondence. One explanation could be that the preformed antibody recognizes denatured antigen *In vivo*. Another explanation is that the isolated peptide adopts a conformation that is able to mimic the surface features of a discontinuous epitope.

The Extended Molecular Self: Human Life as Symbiosis

Consideration of the self necessitates examination of self-immunity, tolerance, and a microbiome-extended or super-self. Tumour cells typify these issues well. Tumours are part self yet can be recognized as non-self by the immune system. In a case where tumor development is associated with the acquisition of gene mutation and expression, immune recognition may beget action (Mapara and Sykes 2004). In this case, the cells of a malignant tumor may be considered as non-self. However, in other cases the tumor's cells are not recognized because of deletional tolerance; most tumour antigens are normal self antigens for which antigen-reactive cells have been eliminated. So what is the general answer to the question: Are cancer cells non-self? The cells are the same cells in both cases. They are cancerous cells that are sometimes being tolerated and sometimes not. This fact cannot be changed – just the meaning associated with each instance. As we can see, there is no simple or categorical answer to the question: "What is the immune self." Nor is there a simple algorithm to answer this question.

Tolerance is the complementary aspect of autoimmunity. It concerns the immune system's ability to ignore its own constituents. The interesting thing is that these constituents may be ignored even if they do not bear the genetic identity marker of the "self." *E. coli* sits in our colon, as well as in the flora of the mouth, without being attacked by our immune system. This bacterium is clearly not a part of the self as defined by the genetic-reductionist approach. How can we explain this tolerance?

Tolerance to parasites is not an exception in nature and *E. coli* is merely a specific instance. Sometimes, intruders to the host develop a unique machinery to hide their own identity. However, in many other cases they are simply tolerated by the host. Organisms, host a variety of parasites that live in perfect symbiosis with them. These parasites, such as *E. coli* are not a part of the self in the genetic sense. The immune system tolerates the presence of the *E. coli*., a fact which the reductionist approach to the immune self may find difficult to explain.

Human symbiosis with our own resident microbes is of the greatest interest and the most pertinent to the present discussion. Throughout life we happily coexist with our indigenous microbial populations of bacteria and also fungi, protozoans and viruses. It is the view of many that human and microbial cells are best thought of as forming a single whole: a composite superorganism known as the "microbiome." Adherents to this paradigm contend that the balanced relationship between the body and its normal flora is crucial to the maintenance of health and proper function. The microbiome undertakes a variety of function roles: control of pathogen growth, they aid the breakdown of food, produce vitamins, confer cross-reactive immunity and aid tissue development. Such an important and exquisite relationship is contingent upon a complexity of interactions which occur both within and between the host, its resident microbial population and the environment. The role of the microbiome in supplementing host defence is most pertinent here (Flower

2008). Stable populations of indigenous microbes provide critical protection against environmental disease-causing pathogens. Since every niche is filled, pathogens cannot out-compete the resident microbes.

The total number of bacteria residing in or on the human body exceeds 100 trillion; equating to 10 per human somatic or germ cell. The total number of different microbial genes in the human body is thought to outnumber human genes by up to 1,000 to one. The human microbial population is composed of an ever escalating number of species, which now number well in excess of 1,000 different species.

This microbiome-centered view of host-microbe interactions has led many popularizers of science to make equivocal and ambiguous statements that some might regard as misleading. Statements such as "are you more microbe than man," have suggested that microbes infest every tissue to the same extent, which is clearly untrue. Muscle, brain, kidneys, fatty tissue and deep lung have long been thought to be sterile, while it is clear that many organs – the intestines, the buccal cavity, the skin and the mucosal lining of the respiratory tract, which sheds 10 000 tiny bacteria-laden water droplets every minute – are home to a significant microbial population, and there is now evidence that many other tissues – bones, blood, joints and arterial cells – support bacterial populations of varying types.

Of all these locations, the gut is by far the most significant. The total bacterial population, weighing about 1.5 kg, coats the 300 m^2 interior lining of the intestine; this population is a complex and dynamic symbiotic living system, and is composed of a large and interacting group of bacterial species. Despite comprizing few divisions, it is very diverse at the level of strain and subspecies, and is able to help protect the host from pathogenic intruders. During our first year, we humans quickly gain significant, stable microbial intestinal communities. Like the guts of ruminants and termites, our distal intestine (which contains most of our gut microbes) is, in effect, a bacterially programmed anaerobic bioreactor which digests a wide variety of otherwise resistant polysaccharides, such as pectin, cellulose and certain starches.

Our gut microbiome is composed of different cell lineages and is able to communicate with itself and with us. It consumes, stores and redistributes energy. It can maintain and repair itself through, and it has co-evolved with its host, complementing and manipulating human biology for the benefit of both host and microbe. The collective genomes of the microbiome supplement our own genetics and metabolism, providing functionality, such as harvesting otherwise inaccessible nutrients, which we have not needed to evolve ourselves. The microbiome may be seen as a microbial organ within the host. We are thus all a composite of not one but many species. We possess our own meta-genomes comprizing genes from our own host genome and from that of our diverse microbial passengers. The microbiome, which may comprise over 100 times the number of host genes, imbues us with many additional functions. The implications for the human immune system are beginning to emerge, enlarging our picture of self versus non-self to include some or all of this microbial population.

Discussion

In this review, we have striven to explore and to examine aspects of the immune self, and its discrimination from non-self. The immune self is often said to be a conceptualization of immune processes, but can also be thought of in terms of molecular patterns recognized by immune agents: epitopes primarily, be they generic epitopes recognized by innate immunity or lipids, carbohydrates, and peptides recognized by T cells and antibodies.

The immune system is a human construct, which is not to say that host immune responses are in any way illusory; they are no mere anthropomorphized canard or phantom: no, they are as real as real can be. The immune system is what stands between us and death from infectious diseases. However, the immune system is a concept rather than a discrete and compartmentalized organ. It arises out of the whole organism of which the host is comprized, and emerges spontaneously at many levels from the molecules and cells of which it is comprized.

The components of the immune system exist at many length-scales: molecules, cells, tissues, and organs. Immunology, as the study of the immune system, is engaged in the exquisite dissection of these components and, through inductive or synthetic reductionism, a proper exploration of its behavior. In time, synthetic reductionism will allow a still elusive predictive understanding of immune processes to emerge from this century long endeavour.

The identification of the agents and objects involved in mediating the immune response is a tractable if laborious task, and a task with many spectacular successes to its credit. That this task is a long way from being completed is a well understood if poorly articulated assertion. However, assigning human descriptions to molecules and cells, such as antibodies and lymphocytes, which reside in their own domain and possess no consciousness, at least as we understand our own psyches, is at best a risky business. We are bound by our partial ignorance not our partial knowledge. An equally difficult yet equally tractable task is charting how these agents work together to make manifest immunity at the whole organism level. However, defining the underlying basis of these behavioral patterns is a task whose difficulty is not to be underestimated. However, the immune self is not a "real" entity as lymphocytes, cytokines or the Thymus can be thought of as real entities. The immune self is not a platonic, autonomous and monolithic entity but a context dependent construct. Self or non-self is not defined by reference to a single specific entity, rather it can be defined as the mutable response by a complex system to a diverse collection of molecular entities.

The boundaries of self and non-self is defined by the recognition properties of the immune system and these can be defined and understood in terms of the physical properties of the molecules composing the immune system. Thus, the properties embodied in the terms "self" and "non-self" are properly emergent. There is no simple positive definition of the immune self as suggested by genetic reductionism. There is no simple negative definition as suggested by Clonal Selection Theory. In a not wholly satisfying way, a practically useful definition of self emerges from the sum of its parts.

It may be that we ultimately find it helpful to consider the terms "self" and "non-self" as physically embodied entities rather than as phantom signals; and that these entities are determined by the mechanisms of the immune system. For non-self, some of these entities are created and then recognized by the immune system (processed and presented T cell and B cell epitopes), and others, such as PAMPS and B cell epitopes, are simply recognized.

References

Joseph J (2002) Twin studies in psychiatry and psychology: Science or pseudoscience? Psychiatr Q 73:71–82

Lykken DT (2006) The mechanism of emergenesis. Genes Brain Behav 5:306–310

Flower DR (2008) Bioinformatics for vaccinology. Wiley Blackwell

Matzinger P (2002) An innate sense of danger. Ann NY Acad Sci 961:341–342

Ada G (2008) The enunciation and impact of Macfarlane Burnet's clonal selection theory of acquired immunity. Immunol Cell Biol 86:116–118

Ferdinand de S (1983) Course in General Linguistics. In: Bally C, Sechehaye A (eds) Trans. Roy Harris, La Salle, IL: Open Court

Jerne NK (1971) The somatic generation of immune recognition. Eur J Immunol 1:1–9

Cohen IR (2007) Biomarkers, self-antigens and the immunological homunculus. J Autoimmun 29:246–249

Janeway CA Jr, Medzhitov R (2002) Innate immune recognition. Annu Rev Immunol 20:197–216

Southan C (2004) Has the yo-yo stopped? An assessment of human protein-coding gene number. Proteomics 4:1712–1726

Nordström KJ, Mirza MA, Larsson TP, Gloriam DE, Fredriksson R, Schiöth HB (2006) Comprehensive comparisons of the current human, mouse, and rat RefSeq, Ensembl, EST, and FANTOM3 datasets: Identification of new human genes with specific tissue expression profile. Biochem Biophys Res Commun 348:1063–1074

Clamp M, Fry B, Kamal M, Xie X, Cuff J, Lin MF, Kellis M, Lindblad-Toh K, Lander ES (2007) Distinguishing protein-coding and noncoding genes in the human genome. Proc Natl Acad Sci USA 104:19428–19433

Levy S, Sutton G, Ng PC, Feuk L, Halpern AL, Walenz BP, Axelrod N, Huang J, Kirkness EF, Denisov G, Lin Y, MacDonald JR, Pang AW, Shago M, Stockwell TB, Tsiamouri A, Bafna V, Bansal V, Kravitz SA, Busam DA, Beeson KY, McIntosh TC, Remington KA, Abril JF, Gill J, Borman J, Rogers YH, Frazier ME, Scherer SW, Strausberg RL, Venter JC (2007) The diploid genome sequence of an individual human. PLoS Biol 5:e254

Doerfler W (2008) In pursuit of the first recognized epigenetic signal – DNA methylation: A 1976 to 2008 synopsis. Epigenetics 3:125–133

Sims RJ III, Reinberg D (2008) Is there a code embedded in proteins that is based on post-translational modifications? Nat Rev Mol Cell Biol 9:815–820

de Diego JL, Gerold G, Zychlinsky A (2007) Sensing, presenting, and regulating PAMPs. Ernst Schering Found Symp Proc 3:83–95

Areschoug T, Gordon S (2008) Pattern recognition receptors and their role in innate immunity: Focus on microbial protein ligands. Contrib Microbiol 15:45–60

Kornbluth RS, Stone GW (2006) Immunostimulatory combinations: Designing the next generation of vaccine adjuvants. J Leukoc Biol 80:1084–1102

Aguilar JC, Rodriguez EG (2007) Vaccine adjuvants revisited. Vaccine 25:3752–3762

Bayry J, Tchilian EZ, Davies MN, Forbes EK, Draper SJ, Kaveri SV, Hill AVS, Kazatchkine MD, Beverley PCL, Flower DR, Tough DF (2008) Targeting regulatory T cells in vaccination: In silico-identified CCR4 antagonists possess adjuvant activity. Proc Natl Acad Sci 105:10221–10226
Flower DR (2003) Towards in silico prediction of immunogenic epitopes. Trends Immunol 24:667–674
Flower DR, Doytchinova IA (2002) Immunoinformatics and the prediction of immunogenicity. Appl Bioinform 1:167–176
Vivona S, Gardy JL, Ramachandran S, Brinkman FS, Raghava GP, Flower DR, Filippini F (2008) Computer-aided biotechnology: From immuno-informatics to reverse vaccinology. Trends Biotechnol 26:190–200
Davies MN, Flower DR (2007) Harnessing bioinformatics to discover new vaccines. Drug Discov Today 12:389–395
Vyas JM, Van der Veen AG, Ploegh HL (2008) The known unknowns of antigen processing and presentation. Nat Rev Immunol 8:607–618
Lin ML, Zhan Y, Villadangos JA, Lew AM (2008) The cell biology of cross-presentation and the role of dendritic cell subsets. Immunol Cell Biol 86:353–362
Loureiro J, Ploegh HL (2006) Antigen presentation and the ubiquitin-proteasome system in host-pathogen interactions. Adv Immunol 92:225–305
Menéndez-Benito V, Neefjes J (2007) Autophagy in MHC class II presentation: Sampling from within. Immunity 26:1–3
Strawbridge AB, Blum JS (2007) Autophagy in MHC class II antigen processing. Curr Opin Immunol 19:87–92
Florence WC, Bhat RK, Joyce S (2008) CD1d-restricted glycolipid antigens: Presentation principles, recognition logic and functional consequences. Expert Rev Mol Med 10:e20
Mori L, De Libero G (2008) Presentation of lipid antigens to T cells. Immunol Lett 117:1–8
Zajonc DM, Kronenberg M (2007) CD1 mediated T cell recognition of glycolipids. Curr Opin Struct Biol 17:521–529
Barral DC, Brenner MB (2007) CD1 antigen presentation: How it works. Nat Rev Immunol 7:929–941
Blythe MJ, Flower DR (2005) Benchmarking B cell epitope prediction: Underperformance of existing methods. Protein Sci 14(1):246–248
Mapara MY, Sykes M (2004) Tolerance and cancer: Mechanisms of tumor evasion and strategies for breaking tolerance. J Clin Oncol 22:1136–1151

A Bioinformatic Platform for a Bayesian, Multiphased, Multilevel Analysis in Immunogenomics

P. Antal, A. Millinghoffer, G. Hullám, G. Hajós, Cs. Szalai, and A. Falus

Introduction

Despite the grand promises of the postgenomic era, such as personalized prevention, diagnosis, drugs, and treatments, the landscape of biomedicine looks more and more complex. It is more and more clear that the fulfillment of these promises for diseases significant in public health requires new solutions for the representation of biomedical knowledge, new approaches to induction for statistical and causal inferences from observations and interventions, and massive computational resources.

Within the biomedical world the two most important responses to this challenge are the mapping and relatively cheap measuring of the genetic variations, such as single nucleotide polymorphisms (SNPs) and copy number variations (CNVs) and the discovery of a new regulatory mechanism based on microRNAs (miRNAs), again with relatively cheap measuring. The recent mapping of the genetic variations has opened a new dimension in the postgenomic research at all phenotypic levels, such as genomic, proteomic, and clinical, and it sparked a series of new studies of genetic association studies (GAS). The new measurements on the one hand might provide a possible solution for the challenges of small sample size and high dimensionality present in high-throughput methods, but on the other hand their utilization requires integrative models and joint analysis of genetic variations, miRNA, gene expression profiles, and clinical quantities (potentially together with proteomic and metabolic profiles).

Besides the potential heterogeneity of the data in a joint SNP–miRNA–mRNA study, the fusion of earlier, separate, and partial measurements and the fusion of knowledge bases are similarly unsolved. Motivated by the ongoing research on the genomics of asthma at the Department of Genetics Cell- and Immunobiology of Semmelweis University Budapest (DGCI), we will focus on GAS in the following context. We assume that

P. Antal (✉)
Department of Measurement and Information Systems, Budapest University of Technology and Economics, Magyar tudosok korutja 2, 1117, Budapest, Hungary

D.R. Flower et al. (eds.), *Bioinformatics for Immunomics*, Immunomics Reviews 3,
DOI 10.1007/978-1-4419-0540-6_11, © Springer Science+Business Media, LLC 2010

1. The goal is the investigation of the genomics of a multifactorial disease using a multi-step, multi-plate partial genome screening (PGS) (e.g., capable for genotyping a hundred SNPs in hundreds of patients)
2. Genotype data sets from PGSs for the same (and other) diseases are available
3. Genotype data sets from genome-wide association studies (GWASs) (and HAPMAP) are available
4. Separate gene expression data are available in the same domain, though possibly in other samples or species
5. Separate microRNA expression data are available in the same domain, though possibly in other samples or species

The chapter is organized as follows. In Sect. 2, we delineate the bioinformatic challenges in connection with the analysis of the genetic background of the multifactorial diseases and our current strategy. In Sects. 3 and 4, we present our exploratory system and decision support system (DSS) for GAS design, which can incorporate diverse prior knowledge and heterogeneous data sets. We compare it against other solutions and the manual approach. In Sect. 5, first we present the results of the standard analysis of a data set of our asthma research. Next, we overview the methodology of Bayesian multilevel analysis (BMLA) and the results of its application. We discuss the integration of these results in the phases of study design, subsequent data analysis, and interpretation. In Sect. 6, we present a methodology for the fusion of large amount of factual domain knowledge and the results of BML analysis. We conclude with discussing the advantages of the Bayesian framework and specifically the Bayesian platform we are developing to cope with challenges of GAS, such as small sample size, fusion of data and knowledge, multiple testing, meta-analysis, and positive results bias. Our study of the genomic background of asthma will serve as a real-world application domain, although the SNPs are anonymized in our bioinformatic illustrations, because the biomedical publications of these results are still in progress.

Bioinformatic Challenge of the Multifactorial Diseases

It is widely accepted that most common diseases are multifactorial, that is, they manifest due to interplay of genetic and nongenetic, or environmental factors (Petretto et al. 2007). These common diseases include, for example, cardiovascular diseases (they are responsible for more than 30% of all deaths in the developed countries), type 2 diabetes mellitus (its prevalence is between 5% and 50% in different populations, and is still increasing), hypertension (its prevalence is 25–35% in developed countries), obesity (20–75%, in developed countries, still increasing), Alzheimer disease (about 5% of the European population, and 20% of the population above 80-year old), or asthma (the estimated number of asthmatic patients in the world is 300 million, and by 2025 it will be 400 million) etc. Considering the high prevalence of these diseases, their economic significance is enormous, for example, in the United States about $100 billion is spent yearly only for cardiovascular diseases, $11 billion for asthma, or in the whole world 2–7% of all health care expenses are spent for obesity-related diseases. Government agencies recognizing that clearing up the genetic background of these diseases will contribute significantly to the development of better (even personalized) treatment, or prevention of the diseases, have spent billions of dollars to search for associated genes or DNA variants in the human genome.. However, in contrast to the monogenic disorders the genetic background of the multifactorial diseases is still poorly understood, the results of the genetic studies are quite disappointing (Shriner et al. 2007).

Earlier, in these investigations the most popular methods were the candidate gene association studies, in which genetic variations in candidate genes, which were assumed to play some role in the diseases, were analyzed. The disadvantage of this method is that it is unsuitable for discovering new metabolic pathways and genes, and because several research groups investigated the same genes, a lot of genetic variations were found which could not been verified in other studies. For example in asthma more than 600 gene association studies were published and more than 120 genes have been found associated with an asthma- or atopy-related phenotype, but only 54 genes have been replicated in 2–5 independent studies, 15 genes in 6–10 independent studies, and 10 genes in >10 independent studies. One reason of this is the several false positive associations. Even if a researcher only tests one SNP for one phenotype, if many other researchers do the same and the nominally significant associations are reported, there will be a problem of false positives (Szalai et al. 2008).

In contrast to the candidate gene studies, the genome-wide screens, the whole or partial genome association studies do not require prior knowledge about the pathomechanism of the disease. The genome-wide screening investigates the whole genome using microsatellite markers through linkage analysis. The hypothesis-independent nature of this approach means that it might more reliably identify susceptibility genes, particularly those in pathways that are not obviously implicated in the disease phenotype. However, genome-wide screens are costly, are labor intensive and can suffer from the lack of statistical power. In addition, when this approach identifies linkage regions, they are generally broad chromosomal regions that contain many genes that could potentially be related with the actual phenotypic process. So, the move from broad linkage regions to causal genes requires additional approaches, including fine mapping and positional cloning of genes within these narrowed regions.

In recent years several companies offered methods for high throughput SNP screenings for a relative cheap prize. The most powerful and promising methods are the GWAS (Petretto et al. 2007). GWAS involve scanning thousands of samples, either as case-control cohorts or in family trios, utilizing hundreds of thousands of SNP markers (there is a chip on the market, which can measure 1 million SNPs) located throughout the human genome. Algorithms are applied that compare the frequencies of either single SNP alleles, genotypes, or multimarker haplotypes between disease and control cohorts. In comparison to family linkage-based approaches, association studies have two key advantages. First, they are able to capitalize on all meiotic recombination events in a population, rather than in a smaller group, like a family. Because of this, association signals are localized to small regions of the chromosome containing only a single to a few genes, enabling rapid detection of the actual disease susceptibility gene. Second, GWAS allow the identification of disease genes with only modest increases in risk, a severe limitation in linkage studies, and the very type of genes one expects for common disorders.

An alternative of the GWAS is the PGS. It utilizes the results of the whole genome screenings, studying genome regions that have previously been found associated with the disease with dense SNP markers. In comparison with GWAS its advantage is that it is cheaper and easier to evaluate. Naturally, it gives significantly less information about the genomic background of the disease in question, but due to the less markers and less questions analysed in one experiment, the statistical evaluation of the PGS is significantly less challenging.

Contrary to the great expectations and billions of dollars spent, the results of GWAS until now have proved to be quite disappointing. For example in asthma (as well as in most other complex diseases) it is widely accepted that several hundred genes and genetic variations must play roles in the pathomechanism of and susceptibility to the disease. But in the only GWAS carried out in asthma so far, measuring 317,000 SNPs in 900 asthmatic and 1200 healthy children (Moffatt et al. 2007) only one gene has been found associated with the disease in the whole genome. The situation is the same in the other diseases, as well. In all cases only one or only a handful of genes have been found. One of the explanations for this is the problem with the multiple testing. This refers to the problem that arises when many null hypotheses are tested; some significant results are likely even if all the hypotheses are false (Balding 2006). The frequentist paradigm of controlling the overall type-1 error rate sets a significance level α (often 5%), and all the tests that the investigator plans to conduct should together generate no more than probability α of a false positive. In complex study designs, which involve, for example, multiple stages and interim analyses, this can be difficult to implement, partly because it was the analysis that was planned by the investigator that matters, not only the analyses that were actually conducted. However, in simple settings the frequentist approach gives a practical prescription: If n SNPs are tested and the tests are approximately independent, the appropriate per-SNP significance level α' should satisfy $\alpha = 1-(1-\alpha')^n$, which leads to the Bonferroni correction $\alpha' \approx \alpha/n$. For example, to achieve $\alpha = 5\%$ over 1 million independent tests means that we must set $\alpha' = 5 \times 10^{-8}$. However, the effective number of independent tests in a genome-wide analysis depends on many factors, including sample size and the test that is carried out. Although the 5% global error rate is widely used in science, it is inappropriately conservative for large-scale SNP-association studies. One of the main features of the multifactorial diseases is the genetic heterogeneity. It means that the increased disease risk is due to several variations with weak effects, and the distribution of these variations is different in different individuals. Variations that can contribute to the increased disease risk in an individual, usually, cannot be found in the majority of other patients, but can also be present in some healthy people. It means that in population level the distribution of the majority of the risk alleles is only slightly different between ill and healthy population. These variations with weak effects cannot be detected with the traditional methods. In addition, the risk alleles usually do not act alone, but in interaction with other alleles and with the environment, further increasing the number of hypotheses tested. Furthermore, there are other measurements, and available additional information that can be utilized when analyzing the genetic background of multifactorial diseases. Variations in the copy number of some sequences (CNVs) are widespread in the genome and can influence significantly the disease risk and the effect of SNPs (Estivill and Armengol 2007). There are also methods and equipment in the market with which high-throughput CNV measurement can be carried out. For example, the Affymetrix chip can measure almost 1 million SNPs and 1 million CNVs at the same time per individual. The CNVs and SNPs also influence the expression level of the genes. In the last years several companies produced equipment and methods which are suitable to measure the expression level of all genes in one measurement in the whole genome generating a huge amount of data. The recently discovered miRNAs play important regulatory roles in the genome and there are also methods available with which their expression levels can be measured (Couzin 2008). The expression level of the miRNAs is influenced by the SNPs, CNVs, the medications, the disease status, etc., and it can influence the susceptibility or the status of a disease (or other phenotype or clinical parameter). Presently there are no available statistical methods that are able to analyze and detect these networks of interactions on a sufficient holistic way satisfying all expectations.

According to the above-mentioned facts it is clear that novel methods and immense computational capacity are badly needed to gain as much information as possible from these genetic studies.

The scientific community and government agencies naturally recognized this need. The National Center for Biotechnology Information (a division of the National Library of Medicine of the NIH) developed the dbGaP [Database of Genotypes and Phenotypes (e.g., diseases)] to archive and distribute the results of studies that have investigated the interaction of genotype and phenotype. dbGaP will serve as the NIH GWAS data repository. Additional information on dbGaP can be found at http://www.ncbi.nlm.nih.gov/sites/entrez?db=gap. The aim of this database is to offer an access to the results of NIH-sponsored GWAS for researchers, and to enable the spreading of new statistical methods once they are developed. Further information can be found in the NIH web pages for GWAS: http://grants.nih.gov/grants/gwas/.

These instances show that there are no available computational methods and capacity in the world at the moment, which are able to analyze these data sufficiently.

The current project of the SNP Core Facility laboratory (http://www.dgci.sote.hu/en/snpcorefacilityen) is a PGS of asthma in asthmatic and control children. The main equipment of the laboratory is the Beckman GenomeLab SNPstream Genotyping System. This system provides an automated scalable system capable of detecting from 4,600 to 3,000,000 SNP genotypes per day and is especially suitable for fine mapping of genome regions found previously in genome-wide screens.

Exploring the Domain

For the rapid tracking of the domain we developed a model-based information retrieval tool and a text mining tool, which are integrated with the study design tool, the Bayesian data analysis tool, and the Bayesian logic interpreter (see Sects. 4, 5, and 6).

Information retrieval (IR) deals with methods for indexing, searching, and recalling data, particularly text and other unstructured data forms. Two major trends leading to an increased efficiency of the IR process are the utilization of user-specific information and domain-specific information. To increase the efficiency of information retrieval in the context of networks, a special language was proposed for the detailed characterization of necessary documents. It allows the definition of complex expressions based on the annotated networks (pathways) in the domain. These expressions are interpreted and evaluated as a relevance measure to select the matching documents by the information retrieval system.

For example, to retrieve documents about a (1) variable "allergy" and (2) its influencing factors (3) in the asthma context (4) with emphasis on "smoking" the following expression can be formulated:

Annotation(Allergy), Annotation[*Neighbors*(Allergy)], Annotation(Asthma), "smoking."

This relevance measure induces the selection of documents that are relevant both to the description of "allergy" and on average to the descriptions of related variables, additionally the presence of the keyword "smoking" is preferred.

Besides the manually curated proprietary knowledge bases such as the Ingenuity system (Ingenuity Systems 2007), to have an instant overview about the publications, we developed a statistical text-mining tool implementing only a shallow linguistic analysis and a multivariate statistical analysis of the literature (see Antal et al. 2004; Antal and Millinghoffer 2006). On the one hand it provides simple statistics about the occurrence, co-occurrence, temporal characteristics of terms (entities) of the domain (see, e.g., Ananiadou and Mcnaught 2006). On the other hand, it performs a Bayesian, multivariate, causally oriented analysis of the literature. The raw and derived bibliometric statistics, the induced models, and causal inferences are available in the study design tool (e.g., for guiding SNP selection), the Bayesian data analysis tool (e.g., for defining informative priors), and the Bayesian logic interpreter (e.g., as derived, high-level factual background knowledge) (see Sects. 4, 5, and 6).

Decision Support System for Design of SNP Association Studies

One of the main problems on the field of SNP association studies is the enormously high number of SNPs and consequently the relatively small sample size. This poses a serious challenge for GWAS and also for PGS, which try to balance cost between the number of variables and the number of samples. At the SNP Core Facility laboratory at the DGCI a main issue is the selection of SNPs before the measurements for PGSs. During the pre-selection it is desirable to select SNPs which are relevant to the specific association study in order to increase the accuracy and power of the statistical analysis.

However, the selection problem cannot be expected to disappear in the near future. While the cost of genome-wide SNP arrays over 1 million SNPs is dropping, the sample size is relatively small. In this case the post-selection is also advisable to increase the statistical power of the data.

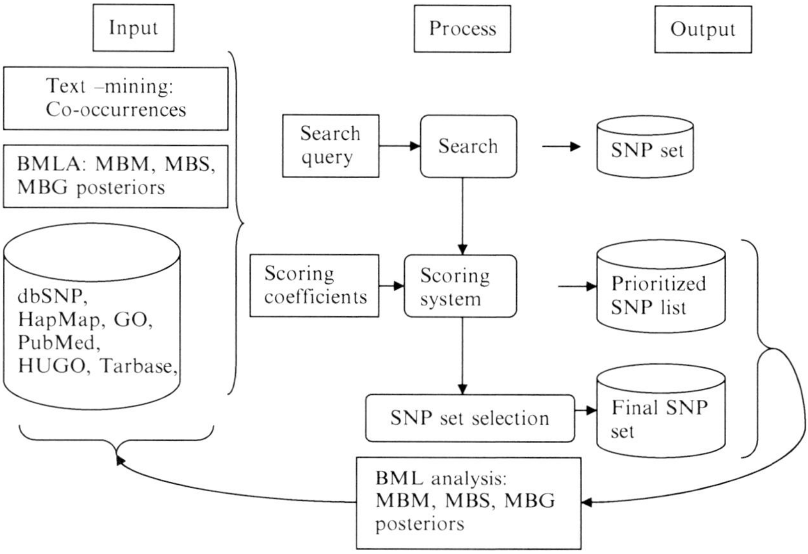

Fig. 1 The figure shows the three steps of SNP selection (see Sect. 4.1). Since the application performs meta-analysis, the input of the program also contains the results of Bayesian statistical analysis while the output of the process provides the basic data set for further analysis

Another important issue in the case of SNP selection before the measurement is to consider technological aspects of the laboratory measurement tools (Beckman Coulter – SNPStream) – whether the SNP set is appropriate for measurement.

To support the pre-selection (for study design) and the post-selection (for data analysis), we decided to follow a normative approach supporting the construction of full-fledged decision network (DN) (Russel and Norvig 2001). Figure 1 shows the links and the functionalities of the DSS.

With the use of the DN it is possible to calculate the biological and medical relevance of each SNP. The DN – using a structured representation – can express the utility or cost of a given SNP set and its relevancy for specific disease.

Because of its clear probabilistic semantic, this formalism supports the easy incorporation of posteriors from earlier studies and it directly provides priors for a subsequent Bayesian analysis as well. To simplify its usage, typical DNs were developed, in which only the parameters have to be modified.

Methods

We collected SNP properties with the help of physicians and biologists from the Semmelweis University to estimate the biological and medical relevance of a SNP set. The presented solution offers a graphical interface to the researchers, wherewith it is possible to give their preferences to score these properties. One of the biggest challenges is to design a scoring system that is easy to use while it is still informative and efficient.

The program uses several databases of the NCBI project such as dbSNP, PubMed, and others like TarBase, miRBase, and HUGO. It performs meta-analysis on results of text and data mining processes, previous measurements, and Bayesian statistical analysis.

The program performs three actions in three different steps controlled by the user:

1. The first step is a search which is designed to replace other SNP searching methods used by the researchers before (UCSC, HapMap)
2. In the course of the second step it is possible for the users to set coefficients to express preferences about SNP properties. In this step with the help of the DN every SNP gets a score expressing its utility and relevance
3. Finally, the third step performs the set selection, the program tries to cover the whole given sequence while chooses relevant and (if necessary) measurable SNPs

In the first step it is possible to set the investigated genome region and also three sets of SNPs:

1. Preferred SNPs – these are SNPs which will be in the final set. Other SNPs correlating with this set get higher scores
2. Stop SNPs 1 – these are SNPs which are probably in the investigated region, but are not important for the researchers. These SNPs will not be in the final set. Other SNPs in the region which correlate with some of these SNPs will get lower score. This set contains for example SNPs which are known not to associate with the specific disease
3. Stop SNPs 2 – these SNPs are somehow important for the researchers, but will not be in the final set. Other SNPs in the region which correlate with some of these SNPs will get higher score. These are SNPs for example which are known to associate with the disease, but already investigated

It is also possible to set a search query in CNF (Conjunctive Normal Form – conjunctions of disjunctions) like: *(MAF>5% OR MBM(SNP, Asthma) OR cooc(SNP, asthma)) AND (genotype=G/C OR SNPFuncion=UTR)*. In the example *MBM* means a pairwise relation according to the Bayesian analysis (see Sect. 6), and *cooc* means a co-occurrence relation based on statistical text analysis (see Sect. 3).

The result of the search is an SNP set in which the SNPs have the appropriate properties, and are found on the given genome region.

In the second step the program scores the SNPs by their biological and medical relevance, this will be the utility of the SNPs. These scores are calculated by the multiplication of the coefficients given by the user (as mentioned before) and the calculated points derived from the data sources.

The scored properties are the followings: (1) The function of the SNPs; (2) The correlation with the three given SNP sets (preferred, STOP 1, STOP 2); (3) If the SNP is on a gene region it gets scores. Genes have scores calculated with the use of GO (http://www.geneontology.org/) and a predefined lexica. (4) The SNP can also get scores if it is related to a micoRNA.

The result of this step is a list of SNPs based on the scores. The members of this list are exactly the same as the members of the search result set.

The third step has three partly controversial goals: (1) to cover the whole region, (2) to choose SNPs with high scores, and (3) to choose measurable SNPs if necessary.

The algorithm uses set selection methods. The result of this step is a set which contains some members of the result list of the previous step. These are the SNPs which are chosen for measurement or further analysis.

Comparing to Other Solutions

In this section we present some applications of this field and compare their properties with the presented application.

The presented application offers a multilayer searching and scoring system to select SNPs for association studies. Its main powers are the DSS, which uses various data sources to process a complex scoring on SNPs; the partial set selection, which selects a SNP set for measurement from the prioritized SNP set; and the integrated measurement technological filter. With the software it is also possible to directly design for more measurements (more plates).

Results

We are still searching for an informative evaluation method. Since the first aim of the program is to decrease the SNP selection time by running processes performed by human resources before, we compared the SNPs chosen by the program with those which were chosen by researchers for a previous measurement. The three investigated regions were the regions 59.500.000–60.500.000 and 61.820.330–62.097.927 on chromosome 11 and the 51.187.000–54.728.000 on chromosome 14. The program chose 90 SNPs to cover every region, 31 of them were also picked up by the researchers manually. The whole SNP selection process for the three genome region was performed in 50 min while this job takes several weeks for the researchers.

Known Issues and Future Goals

The program loses approximately the half of the SNPs during the computation, since in the dbSNP database the information about SNPs is incomplete while the researchers work with this incomplete data set. Another open problem is that the application is designed to select an SNP set for one or more plates from a specific region, but the researchers used to choose SNPs for one plate from more than one genome region. Finally, in general we want to use more data sources for the scoring system, particularly more results from our text-mining method and manually curated knowledge bases (see Sect. 3), and results from the statistical analysis of earlier measurements (see Sect. 5).

Data Analysis

The new high-throughput measurement methods in biomedicine provide a massive amount of data such as SNPs, CNVs and comparative genome hybridization, gene expression, micro RNA, or proteomic profile. In the predictive context the target variables are called the response, outcome, class, or dependent variables and the rest of the variables are called features or predictor, explanatory, or independent variables (or even factors and covariates). In our biomedical domains the target variables are the status, stage, or grade of the disease and the features are the familial and personal descriptors, the clinical symptoms, and the genomic and proteomic variables. The selection of the relevant variables in the predictive context is called the feature subset selection (FSS) problem (Kohavi and John 1997).

A Bayesian Primer

From a practical point of view the requirements for a Bayesian statistical analysis are similar to the standard ("frequentist" or hypothesis testing) statistical framework (FSF), but the results open up completely new ways of thinking and communication that are essential in the post-genomic era. To overview some concepts of Bayesian decision theory and the Bayesian statistical framework (BSF), for simplicity, we assume discrete models M_i. The crucial and fiercely debated concept of BSF is the prior distribution (prior) $p(M_i)$ representing the subjective beliefs in a given context. Using the marginal model likelihood or evidence $p(D|M_i)$ for a given data set D and model M_i, the famous Bayes rule gives the posterior distribution (posterior)

$$p(M_i \mid D) = \frac{p(M_i)p(D \mid M_i)}{p(D)} \propto p(M_i)p(D \mid M_i), \tag{1}$$

which represents beliefs in model M_i after observing data set D. This equation also shows that the posterior is a kind of equilibrium between the prior and the likelihood, and with an increasing number of observations, the posterior is more and more dominated by the likelihood (data) and the effect of prior becomes negligible.

Neglecting the possibility of incorporating valuable prior information using the prior, the distinctive feature of the BSF is its direct statement about the plausibility of a given model (cf. the hypothesis testing framework). Besides the clear interpretation, the directness of the statistical conclusions allows their arbitrary postprocessing, combination, and fusion. Assuming that the given model property (or feature) F is important in the analysis, we can induce a posterior expressing its plausibility as follows:

$$p(F|D)=\Sigma_{M_i:\text{"}F \text{ is true in } M_i\text{"}}\, p(M_i|D). \quad (2)$$

The operation of integration or summation over models is termed in this context as Bayesian model averaging. Usually these cannot be performed analytically, thus stochastic simulation methods can be used to approximate the inference (see, e.g., Gelman et al. 1995).

Assuming a utility (or loss) function for the selection (e.g., report) of model M in case of M_0, $U(M,M_0)$, the selection of the optimal model M^* with minimal loss is dictated by the Bayesian decision theory as follows

$$M^* = \text{argmax}_{M_i} \sum_{M_0} U(M_i, M_0)p(M_0 \mid D) \quad (3)$$

Next, we compare the main properties of the frequentist and BSFs, particularly relevant in bioinformatics applications:

- *Direct probabilities of hypotheses.* From setting up the model to the final inference uncertainties are expressed exclusively in a single coherent system of probabilities, specifically the statistical inferences are direct statements about the probabilities of hypotheses (cf. the refutation of a hypothesis at a given significance level to reach a statistical conclusion in the FSF).
- *Small sample applicability.* Because the inferences have a direct, probabilistic interpretation, Bayesian methods can be applied irrespectively of the sample size.
- *Results conditioned on all data.* Because the inferences have a direct and joint interpretation, the remaining uncertainty can be normatively interpreted. Thus resampling techniques are not necessary for a robust inference as in FSF, consequently the complete data set can be used in the statistical inference.
- *Priors.* Bayesian statistics can be knowledge rich (intensive), because the prior distribution allows the incorporation of background domain knowledge (cf. in FSF only the selection of the model class can be used for such purpose).
- *Complex models.* Complex models can be used because of the inherent model averaging.
- *Bayesian decision theory.* The direct probabilistic predictions and inferences allow the immediate application of the Bayesian decision theoretic framework and the use of the conclusions as priors in subsequent statistical analysis.
- *Harnessing computational power.* The application of Bayesian statistics is typically computer intensive, thus it can exploit and motivate the development of high-performance computations. This is the consequence of the fact that the central operation of Bayesian statistics is model averaging, which is more costly than model selection (i.e., optimization) in the frequentist framework.
- *Solutions for the multiple testing problem.* Complex models and Bayesian model averaging – possibly further combined with resampling techniques – provide a solution for the multiple testing problem, a serious concern in GASs.

Basically, the use of a complex model and Bayesian model averaging provides a normative, self-calibrating characterization for the remaining uncertainty (i.e., for the power or sufficiency of a given data set).

- *Combination of analyses.* Because the inferences have a direct and joint probabilistic interpretation, the inferences can be combined transparently and directly. Consequently, the off-line approximations of the posteriors of various statistical analyses can be treated and fusioned as single probabilistic knowledge-base.

A Bayesian Network Primer

Bayesian networks form a subclass of graphical models that is using directed acyclic graphs (DAGs) instead of more general graphs to represent a probability distribution and optionally the causal structure of the domain (e.g., see Fig. 2). In an intuitive causal interpretation, the nodes represent the uncertain quantities, the edges denote direct causal influences, defining the model structure. A local probabilistic model is attached to each node to quantify the stochastic effect of its parents (causes). The descriptors of the local models give the model parameters. The widespread popularity of this representation is probably the consequence of its applicability in multiple disciplines. The multifaceted nature of Bayesian networks follows from the fact that this representation addresses jointly three autonomous levels of the domain: The causal model, the probabilistic dependency–independency structure, and the distribution over the uncertain quantities.

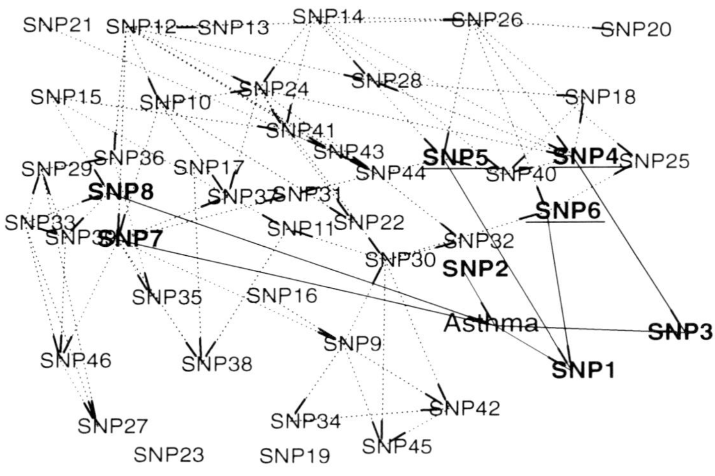

Fig. 2 The relevant variables, their interactions, and the maximum a posteriori complete Bayesian network model of the domain. The MAP MBS set containing the relevant SNPs for asthma is denoted by bold symbols, The MAP MBG representing their interactions are denoted by solid lines. The conditionally relevant variables (only appearing in interactions) are denoted by underscore

Bayesian Network Properties for Representing Relevance

The standard probabilistic definition of association (relevance) is as follows. A set of variables X' is called a Markov blanket (MBS) of X_i in a given domain with variables $X_1,...,X_n$, if X_i is independent of the rest of the variables given the set X' (Pearl 1988). Bayesian networks and BN properties (features) offer a wide range of options for representing relevance. For a distribution *P* defined by Bayesian network *G* the "neighbors" form a Markov blanket of X_i, where the "neighbors" are defined as the set of parents, children, and the children's other parents for X_i (Pearl 1988). The parent–child relation in directed graphs is as follows: "Parent→child." The induced (symmetric) pairwise relation $MBM(X_i, X_j, G)$ is defined as X_i and X_j are "neighbors" in *G*. See Fig. 2 for the illustration of MBS and MBM concepts.

To exactly cover the interactions of the relevant variables – besides their multivariate exploration – we proposed the use of Markov Blanket (sub)Graph (MBG) feature (property), a.k.a. classification subgraph (Antal et al., 2006). A subgraph of Bayesian network structure *G* is called the Markov Blanket (sub)Graph or Mechanism Boundary (sub)Graph $MBG(X_i,G)$ of variable X_i if it includes the "neighbors" in G and the incoming edges into X_i and into its children. For a probabilistic and causal interpretation, see Antal et al., (2006). An important property of the MBG feature is that it is sufficient for the relevance analysis of X_i.

The Markov Blanket sets and Markov blanket subgraphs reflect a new approach to the use of properties of complex models and Bayesian model averaging. The earlier approach tries to provide an overall characterization as a fragmentary representation (Friedman and Koller 2003). Such features are pairwise edges, compelled edges, and Markov blanket relations. At the other extreme of feature learning we find the identification of arbitrary subgraphs with statistical significance (Peer et al. 2001). This is close to our approach to Bayesian network features studied in this chapter, but we restrict the subgraphs to Markov blanket subgraphs to have a focused representation from a single, but complex point of view (i.e., from the point of view of the selection of relevant variables) and we use the Bayesian framework instead of the frequentist framework.

The Bayesian approach to the FSS problem leaves the question of the model class open, especially the use of domain models or conditional models. For simplicity, we compare these options as Bayesian networks against logistic regression though more advanced non-parametric Bayesian methods are also attractive options. The advantage of using Bayesian networks is that (1) a closed-form for the structure posterior is usually available (Cooper and Herskovits 1992), contrary to typical conditional models (Denison et al. 2002), (2) there is no need for an auxiliary model to handle missing values, and (3) the parameter and structure distributions are interpretable. On the other hand, conditional models are statistically and computationally less complex than Bayesian networks, and they handle continuous variables and contextual dependencies more generally (Boutilier et al. 1996) [i.e., the properties (1) and (2) are lost in general BNs]. In this chapter, however, we will focus on the use of BNs [for a different approach, see e.g., (Zhang and Liu 2007)].

The Bayesian Multilevel Data Analysis

The goal of the Bayesian multilevel analysis of relevance is to calculate and crosslink the posteriors corresponding to features X_i, sets of features, (sub) graphical models of features, and a target variable Y. Following our assumption in this chapter about the underlying BN representation, it means the posteriors for the Markov Blanket Memberships ($MBM(Y, X_i)$), Markov Blanket sets ($MB(Y,X')$), and Markov Blanket graphs ($MBG(Y,MBG)$). Further levels are also possible either using domain specific knowledge for defining groups of variables w.r.t. their types, see Sect. 6, Table 3.

At each level we report the most probable feature values for interpretation and for comparison with frequentist results. Additionally, to demonstrate the applicability of the Bayesian approach to characterize the remaining uncertainty (i.e., the relative power of the data), we report indicators for the peakness of its posterior. As another important indicator for the sufficiency of the data is the computation of the posteriors for varying sample size, we report both learning curves and sequential analysis using the temporal ordering of the cases.

To estimate the posteriors we applied a DAG-based and an ordering-based MCMC method [for overviews and related works on MC for BN features see, e.g., (Friedman 2003). The DAG-based method uses a Metropolis Coupled

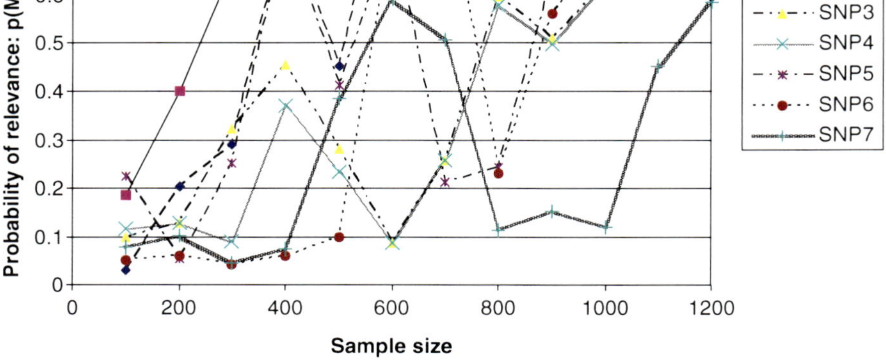

Fig. 3 The sequential posteriors – inferred from growing amount of data – that a given SNP is relevant for asthma. The probability of relevance is represented by the posterior $p(MBM(X_i,Asthma|D)$

Markov Chain Monte Carlo technique [see, e.g., (Zhang and Liu 2007)] with a proposal distribution from Castello et al. (Giudici and Castelo 2003). The ordering-based MCMC uses the prior and proposal distributions from Friedman et al. (Friedman 2003) and utilizes the fact that there is closed-form for the ordering conditional posterior of the MBGs (Antal et al., 2006). The length of the burn-in was selected using Geweke's z-score test and the R value of the multiple-chain method of Gelman-Rubin (Gamerman 1997; Gelman et al. 1995). The length of the MCMC simulation was selected to decrease the variances of the MCMC estimates below 0.01.

Table 1 Overview of applications for SNP selection in GAS design

Application	Databases	Decision support	Simple search	Scoring	Set selection	Multilayer	Target field
Genomizer (Franke et al. 2006)	Multiple	No	Yes	Yes	No	No	SNP
QuickSNP (Deepak et al. 2007)	One	No	Yes	No	No	No	SNP
GoldSurfer (Fredrik et al. 2004)	One	No	Yes	No	No	No	SNP
HAPLOT (Sheng et al. 2005)	One	No	Yes	No	No	No	SNP
SNPHunter (Lin et al. 2005)	One	No	Yes	No	Yes	No	SNP
Endeavour (Stein et al. 2006)	Multiple	Yes	Yes	No	Yes	No	Gene
SNPSelector (Hong et al. 2005)	One	No	Yes	Yes	No	No	SNP
OSIRIS (Julio et al. 2006)	Multiple	Yes	Yes	No	No	No	Gene
SNPbrowser (De La Vega et al. 2006)	Multiple	No	Yes	No	No	No	SNP
SNPs3D (Yue et al. 2006)	Multiple	Yes	Yes	No	No	No	Gene/SNP
Presented application	Multiple	Yes	Yes	Yes	Yes	Yes	SNP

Meaning of the columns: Databases – whether the program uses one or more databases. Decision Support – whether the program offers decision support. Simple search – whether the program performs simple search methods. Scoring – whether the program scores the target object to prioritize them. Set selection – whether the program supports the selection of a specific subset of the target elements with decision support methods. Multilayer – whether the solution uses a multilayer system hierarchy. Target field – the type of target elements

Results

We demonstrate the results of the analysis of a data set (containing 46 SNPs selected from the asthma susceptibility region of chromosome 11q13 and 1,175 genomic DNA samples from asthmatic and control patients) from the biomedical domain of asthma. The data analysis process consisted of three parts. First, a set of standard statistical methods was applied on the examined data set. This included the standard test for genetic association (i.e., Hardy-Weinberg equilibrium test), odds ratio computation, Cochran-Armitage trend test, and logistic regression. In four cases the HWE test applied on *control samples* indicated a significant deviation from the expected distribution of the controls with a *p*-value less than 0.001. Since this deviation is a sign of a sampling or genotyping errors, these SNPs were excluded from further analysis. Second, the data set was subjected to BMLA, and finally the robustness of the different methods was analyzed. The results are summarized in Fig. 5 and Table 2, which consist of three parts.

The first part contains the results of the standard statistical methods. Only significant results with a *p*-value less than 0.05 are shown, these are denoted by "1." The *HWE* column shows the results of an HWE test computed on *cases* (i.e., the subset of samples in which the patients were diagnosed with asthma disease). The *ODDS* column presents the significant odds ratios, while the *ARM* column presents the results of the Cochran-Armitage trend test. The next two columns demonstrate the results of the forward likelihood ratio-based logistic regression method. *LRCO* denotes the continuous case (i.e., all SNP variables are treated as continuous), while *LRCAT* denotes the categorical case (i.e., all SNP variables are treated as categorical).

The second part of the table shows the outputs of Bayesian methods. *MBM* denotes the Markov-Blanket Membership probability, that is, the probability that a given SNP is a member of the Markov Blanket of the target variable (*Asthma*). This is a pairwise relationship, where a high probability indicates a relevant association with the target variable. The following columns *MBS-1–9* stand for the most probable Markov Blanket Sets. For each set the corresponding members are marked (with "1"). Being a member of such an MBS indicates that the variable is directly relevant; its effect cannot be blocked by other variables or in other terms its effect is not mediated by other variables. The *MBG-1–6* (Markov Blanket Graph) columns show an even more refined picture. They not only identify the relevant variables, but also the type of interaction between them. Having the target variable as a basis, every member of an MBG can be classified into one of the following types (see Fig. 2):

- SP: Single parent. The single parents *SNP2*, *SNP7*, and *SNP8* have a joint effect on the target variable (i.e., on asthma), indicating their joint presence in a biomedical mechanism. Because they have no direct influence on other relevant variables represented by directed edges, their joint effect statistically, and possibly casually, is separated from the rest of relevant variables (i.e., from the groups of *SNP1-SNP5-SNP6* and *SNP3-SNP4*).

Table 2 Prediction of the relevance of SNPs, sets of SNPs, and their interactions in asthma

	Standard statistical methods					Bayesian statistical methods																		Robustness analysis						Reference sets		
SNP	HWE	ODDS	ARM	LRCO	LRCAT	MBM	MBS-1	MBS-2	MBS-3	MBS-4	MBS-5	MBS-6	MBS-7	MBS-8	MBS-9	MBG-1	MBG-2	MBG-3	MBG-4	MBG-5	MBG-6	DAG-MBM	DAG-MBS	RO-RAN	RO-SET	RO-ALL	RO-HWE	RO-ODDS	RO-ARM	Hi	Med	Lo
SNP1		1		1	1	0.999566	1	1	1	1	1	1	1	1	1	imc	imc	imc	imc	imc	imc	1.00E+000	1	1	1	1				1	1	1
SNP2				1	1	0.941736	1	1	1	1	1	1	1	1	1	sp	sp	sp	sp	sp	sp	8.78E−001	1	1	1	1				1	1	1
SNP3					1	0.918699	1	1	1	1	1	1	1	1	1	imc	imc	imc	imc	imc	imc	9.63E−001	1	1						1	1	1
SNP4	1					0.917069	1	1	1	1	1	1	1	1	1	op	op	op	op	op	op	9.61E−001	1	1			1			1	1	1
SNP5	1					0.848814	1	1	1	1	1	1	1	1	1	op	op	op	op	op	op	3.33E−001		1							1	1
SNP6		1	1			0.845424	1	1	1	1	1	1	1	1	1	op	op	op	op	op	op	3.34E−001		1			1				1	1
SNP7	#					0.55041	1	1	1	1	1	1	1	1	1	sp	sp	sp	sp	sp	sp	2.03E−001		1							1	1
SNP8						0.25951	1	1		1	1	1		1	1	sp	sp		sp	sp	sp	1.06E−001		1							1	1
SNP9		1				0.172314																6.96E−002		1							1	1
SNP10						0.0979096																8.27E−002									1	1
SNP11		1	1	1	1	0.0878048																1.68E−001		1	1	1					1	1
SNP12		1	1			0.0544094																3.28E−002		1	1	1					1	1
SNP13						0.0470249																4.96E−002										1
SNP14		1	1			0.0323594		1					1	1			sc	sc			sc	3.33E−002										1
SNP15						0.0311053																6.57E−002										1
SNP16						0.0258237						1								sc		1.61E−002										1
SNP17	#					0.0217918					1			1					sc		sc	6.87E−002					1					1
SNP18		1			1	0.0138562																2.38E−002										1
SNP19						0.0135745																2.96E−002			1							1
SNP20						0.0134758																3.07E−002										1
SNP21	1					0.011549									1							2.07E−002										1
SNP22						0.0104153																1.27E−001						1				1

(continued)

SNP	Standard statistical methods					Bayesian statistical methods																		Robustness analysis						Reference sets		
	HWE	ODDS	ARM	LRCO	LRCAT	MBM	MBS-1	MBS-2	MBS-3	MBS-4	MBS-5	MBS-6	MBS-7	MBS-8	MBS-9	MBG-1	MBG-2	MBG-3	MBG-4	MBG-5	MBG-6	DAG-MBM	DAG-MBS	RO-RAN	RO-SET	RO-ALL	RO-HWE	RO-ODDS	RO-ARM	Hi	Med	Lo
SNP23						0.0102574				1												2.44E−002										1
SNP24		1	1			0.00898338																2.77E−003										1
SNP25						0.00767404																7.96E−003										1
SNP26		1				0.00563429																6.42E−003										1
SNP27						0.00502285																1.08E−002										
SNP28						0.00426429																4.02E−003										
SNP29	1	1		1		0.0029363																2.03E−003										
SNP30	#					0.00241169																3.47E−003										
SNP31						0.00133164																4.36E−002										
SNP32		1				0.00131282																7.24E−003										
SNP33						0.000578337																1.40E−003										
SNP34						0.000561699																5.53E−004										
SNP35						0.000423999																3.32E−004										
SNP36						0.00041027																0.00E+000										
SNP37						0.000338769																2.00E−003										
SNP38	#					0.00027036																0.00E+000					1					

SNP39	0.000259799	1.50E−003
SNP40	0.000230419	1.79E−003
SNP41	0.000152016	2.56E−005
SNP42	0.000127294	4.62E−004
SNP43	9.75E−05	0.00E+000
SNP44	6.73E−05	0.00E+000
SNP45	4.28E−05	0.00E+000
SNP46	6.04E−06	0.00E+000

HWE – Hardy-Weinberg equilibrium test, *ODDS* – odds ratio, *ARM* – Cochran-Armitage trend test, *LRCO* – logistic regression (continuous case), *LRCAT* – logistic regression (categorical case), *MBM* – Markov-Blanket Membership probability, *MBS-1–9* – most probable Markov Blanket Sets, *MBG-1–6* – most probable Markov Blanket Graphs, *DAG-MBM* – Markov-Blanket Membership probability (DAG-MCMC based method), *DAG-MBS* – best Markov Blanket Set (DAG-MCMC based method), *RO-RAN*, *RO-SET*, *RO-ALL*, *RO-HWE*, *RO-ODDS*, *RO-ARM* – various robustness tests. Highlighted SNPs have failed the HWE test applied on control samples and therefore were excluded from further analysis

- SC: Single child. A single child of the target variable indicates a separate stochastic effect, and possibly a distinct causal mechanism. The *SNP14* in *MBG-2* is such a variable, see Table 2.
- MC: Child with multiple parents and with 0 "other parent." Such a group of variables has a separate stochastic effect, and it possibly indicates a distinct causal mechanism involving the interaction of multiple, participating variable, for example, *SNP1-SNP5-SNP6* and *SNP3-SNP4*.
- IMC: Child with multiple parents and with 0<"other parent." Such a child as *SNP1* and *SNP3* is interesting, because if it is unknown, the knowledge (i.e., observation) of another variable is irrelevant (as *SNP5-SNP6* and *SNP4*).
- OP: Other parent [i.e., another parent of the target node's child(ren)]. This is the opposite of the above. Such parents as *SNP5, SNP6*, and *SNP4* are interesting, because they are irrelevant unless another variable is known (as *SNP1* and *SNP3*).

The latter three cases can be interpreted as interactions (e.g., their effect can be modeled in logistic regression by interaction terms).

The *DAG-MBM* and *DAG-MBS* results are similar to the previously discussed MBM and MBS results. The difference lies within the way of implementation.

The third section of the table contains the results of tests conducted to measure the robustness of both Bayesian and standard statistical methods. *RO-RAN* denotes the case when the samples of the data set were randomly reordered and the data set was cut into half. The MBM probabilities were calculated for the whole data set and for the half data set. Both result sets were examined and every SNP, whose MBM probability was above a threshold of 0.05, was marked as relevant. If an SNP was marked relevant in both the whole and the half data set then it was marked as robust denoted by "1" in the *RO-RAN* column. In case of the *RO-SET* the process was similar except that the data set was divided according to the two subpopulations from which the SNPs were sampled; the two half sets were examined independently. An SNP was marked as robust if the MBM probability was relevant in both data sets. The next column *RO-ALL* is an AND combination of the abovementioned *RO-RAN* and *RO-SET.* Finally, RO-HWE, RO-ODDS, and RO-ARM present the robustness of the standard statistical tests. All three of them were applied on both half data sets that were created according to the two subpopulations. When an SNP was marked significant by a test in both data sets separately, then it was identified as robust in the corresponding RO-HWE, RO-ODDS, RO-ARM column.

Discussion

Although the biomedical interpretation, evaluation, and verification with further measurements of the results of the BMLA method w.r.t. other methods is not yet finished, it has been already proved to be useful for guiding the classical data analysis and for prompting new research and development, particularly because of its model-based, multivariate nature. Specifically, we can summarize our experiences as follows.

At the pairwise level, the advantage of the MBM posteriors compared to the standard pairwise statistical association tests is that despite its pairwise representation it correctly indicates that the variables are part of a multivariate relation. This is probably the consequence that it is still a model-based statistical relation (derived by Bayesian model averaging). Another general advantage of the MBM "scores" are that, as posteriors, they can be visualized with a clear semantics, their multiple use is not hindered by the problem of multiple testing (not even in a sequential analysis with thousands of applications), and their joint usage can be used to characterize the power of the data.

To evaluate the necessity of the multivariate approach, we have to investigate the performance of the MBM scores (posteriors) and the MB scores (posteriors). Note that because of the Bayesian foundation the scores are posteriors, and the MBM posteriors are marginal distributions of the MB posterior. Figure 4 reports the posteriors of the maximum a posteriori (MAP) MB sets and their pairwise MBM-based approximated values. The monotone decreasing curve corresponds to the ranked MB posteriors using the ordering-MCMC.

Good approximation indicates the conditional independence of the features (or the presence of independent mechanisms in a causal reading). As Fig. 4 shows this is not the case with the current data set, as the MBM-based approximation performs badly both w.r.t. estimations and ranks. A consequence of this is that certain SNPs marked as relevant by the multivariate methods (e.g., LRCAT, MBS-x) have low MBM score.

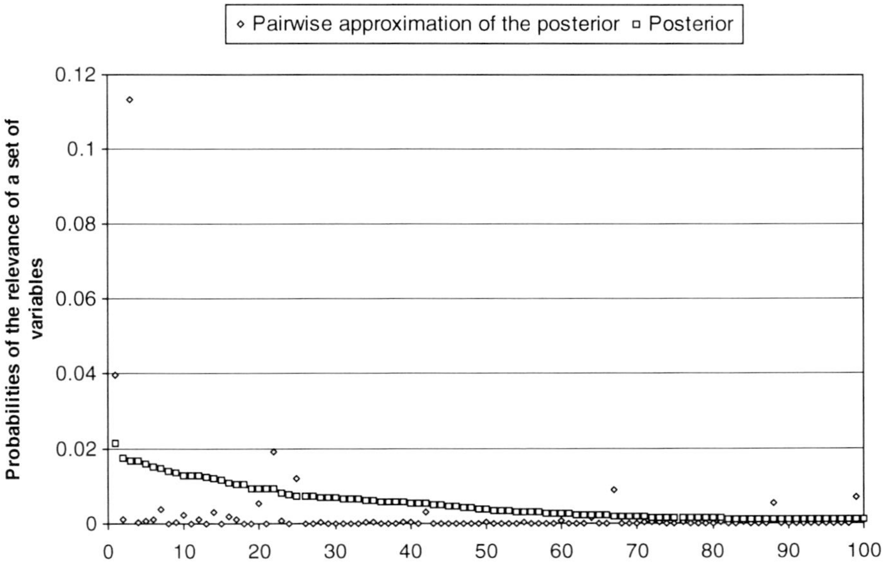

Fig. 4 The peakness of the posteriors of the most probable MB sets and their pairwise MBM-based approximation. The horizontal axis shows the posteriors, the vertical axis corresponds to the appropriate ranks

However, Fig. 4 indicates the lack of dominating MB set, that is, the lack of single set of variables with an outstanding posterior (MB score). The joint evaluation of these indicators at multiple levels supports a detailed evaluation of this phenomenon as follows. Flat posteriors at the MBM and MB levels can be the consequence of insufficient data or redundancy of the features. This question can be answered by the evaluation of the posterior at the MB level, because it explicitly shows the alternative sets (e.g., alternating pairs) of features. Because such alternating sets are not present in the most probable MB sets, probably the data set do not define a posterior at the MB level peaked enough. This is also corroborated by the investigation of the change of the entropy at the MBM and MB levels for growing sample size. Despite the decline of the sum of the entropy corresponding to the MBM posteriors from 20.66 at sample size 100 to 7.59 at all samples, the entropy corresponding to the MB posterior is much less effected (it is changing from 8.71 to 7.08).

Despite the lack of a dominant relevant set, the relevance of certain subsets is quite certain. This brings up the issue of a more detailed analysis of the sets, that is., the investigation of their internal dependencies, which is exactly and compactly performed at the MBG level. The MBG level w.r.t. the MBM and MB levels offers the following properties. First assuming causal prior or interventionist data, it identifies all the multivariate mechanisms related to the target variable (hence its second name Mechanism Boundary subGraph). Second, MBGs indicate conditional relevance explicitly, that is, a variable is not relevant unless others are known. Third, MBGs can be used to indicate the substitutability of a feature, which is an important issue for expensive or clinically, psychologically, not preferred features.

An interesting example for the use of the multivariate approaches and the multivariate-interactionist approaches is the following (see Fig. 5 and Table 2). The SNP4 is not relevant by the odds tests, and standard LR tests, only HWE indicates modest significance (p-value 0.02365). However, this variable is present in the MAP MBG as conditionally relevant variable, with condition SNP3, and the interaction terms of SNP4* SNP3 are indicated as relevant by the odds tests (with p-value $<<0.001$) and by the logistic regression (Beta parameters 0.920, 1.029, 0.347, 1.258). The SNP5, SNP6 are similarly conditionally relevant with conditional term SNP1.

Bayesian Logic for the Fusion of Knowledge and Data

The accumulation of electronically accessible knowledge has a significant impact on study design and statistical data analysis, as we discussed in Sects. 4 and 5. However, it can also be utilized for the "interpretational" bottleneck, that is, to cope with the challenge of the interpretation of statistical data analysis of high-throughput methods. Many methods are investigated, for example, the literature-based keyword profiling of the results of the data analysis or the univariate representation of the results of the data analysis and its representation in the "pathways" of knowledge (Ingenuity Systems 2007). In the following we summarize and illustrate a general method for the fusion of voluminous factual knowledge and the results of statistical

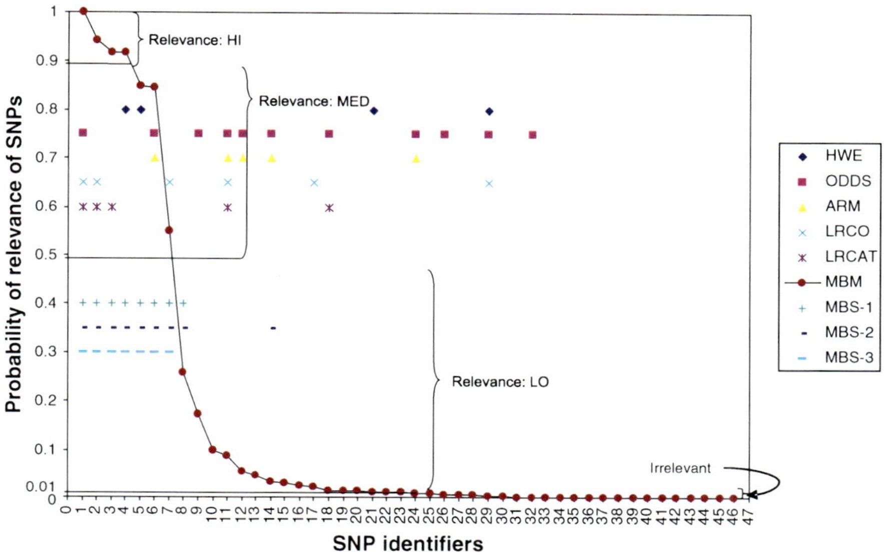

Fig. 5 The overview of the results of data analysis. The notations are as follows: *HWE* – Hardy-Weinberg equilibrium test, *ODDS* – odds ratio, *ARM* – Cochran-Armitage trend test, *LRCO* – logistic regression (continuous case), *LRCAT* – logistic regression (categorical case), *MBM* – Bayesian pairwise relevance, *MBS-1–9* relevant sets by Bayesian analysis. (only MBM values are numeric, others are arbitrary values for visualization)

data analysis, which on the one hand is able to preserve the semantic richness of the knowledge and on the other hand, the multivariate or even model-based foundation of the statistical data analysis.

Factual Sources

A factual source may be anything storing information about the research field in question, typical examples include

- *Expertise*. Information gained from human experts of the domain. Though this knowledge is usually difficult to formalize, it is almost always exploited in the model construction process (e.g., determining the set of variables and their value ranges).
- Textual (kernel) descriptions of variables, keyword, and synonym lists.
- Ontologies string basic data of domain entities and their relations. Such as can be, for example, dbSNP (http://www.ncbi.nlm.nih.gov/projects/SNP), HAPMAP (http://www.hapmap.org), Gene Ontology (http://www.geneontology.org), and Ingenuity (http://www.ingenuity.com).
- Publication repositories, like PubMed (http://www.ncbi.nlm.nih.gov/pubmed).

The Hybrid Knowledge Base

Because the variables of the data analysis (in the "data world") and of the factual data and knowledge bases are the same, we can conceive their merge as a hybrid knowledge base. Earlier we proposed the following components and functionalities:

- Connections between the elements of the underlying Bayesian networks (probabilistic knowledge base) and of the factual knowledge sources (logical knowledge base).
- The syntax of the query language allows any formulas composed of the (structural) properties of Bayesian networks and of the native features of the incorporated factual sources. The semantics of sentences is defined as the probability that they are true.
- An algorithm based on the MCMC sampling of the model space was developed to evaluate queries.

The proposed method and implemented system on the one hand allows the logical combination of the relevant parts of the voluminous factual knowledge sources, and on the other hand it defines and computes a probability for complex statements (hypotheses) based on the results of the Bayesian data analysis. For example the combination of different knowledge sources may allow formulating queries like:

"What is the probability that the SNPs relevant in the probabilistic sense to the target are related to such genes that occur in the same pathway in a given knowledge base?"

The above example could be "translated" as follows:

- Relevant SNPs are those variables that are in the Markov blanket set of the target.
- The SNP-gene assignment can be given by a separate resource (e.g., dbSNP).
- The pathways for the genes can be queried from the prespecified knowledge base.

As an illustration of this probabilistic knowledge base we computed the induced posterior probability of the type of the SNPs and the induced posterior for effected genes. The first target is the probability that a given keyword of SNP types is present in the annotations of the relevant SNPs for Asthma. Table 3 shows the ratio of the types of the measured SNPs, the conditional probability of relevance for a given SNP type [i.e., *P(X0MBS(Asthma)|Type(X) = type)*], and the ratio of types of relevant SNPs [i.e., *P(Type(X) = type|X0MBS(Asthma))*]. The second quantity can be interesting in study design (i.e., the probability of relevance for various types). Note that this is independent of the ratio of the SNPs in the current study. The third quantity is interesting in interpretation. Note that it is a relative quantity with respect to the ratios of the measures SNPs, so their change is interesting, for example, 45.65% of the SNPs measure are in introns, and 61.9% of the SNPs found relevant for asthma are in introns.

The other quantity derived by the Bayesian averaging is the probability that a given gene is present in the annotations of the relevant SNPs for asthma. Figure 6 presents the induced posteriors for the genes assigned to the examined SNPs for growing sample size. Compared to the curves of the posteriors of relevance for SNPS, we can observe a sharper rise for the relevant genes (i.e., faster convergence or larger statistical power), which in general can be expected as the gene level is an aggregated level.

Conclusion

The recent stage of the GAS world is similar to the stage of the gene expression world 5–10 years ago with some marked differences, which can be used to indicate certain trends for the bioinformatics applications. The most obvious, but profound, difference is that the nature of genotype data is more "accumulative," than that of the expression data, corresponding mostly to a person, not to a representative tissue. This accumulative effect will be immanent in the near future for genotyping centers performing more and more PGS studies (irrespectively of the hardly predictable future of private, mass genotyping). Consequently, the analogues of Gene Expression Omnibus (http://www.ncbi.nlm.nih.gov/geo/) such as dbGaP have to be prepared for the gradual accumulation of both genotype and clinical environmental data as well. Thus it can be expected that the use of earlier data sets in meta-analysis will be even more significant. Another trivial difference is that as genetic variations can influence and explain aspects of expression data, databases of genetic variations will probably become the new foci and foundational layer for expression repositories. Furthermore, due to the increasing heterogeneity and sheer volume, several issues, such as the small sample size, the fusion of data and knowledge, the challenges of multiple testing, meta-analysis, and the positive results bias, will be even more pressing.

In our opinion, as a response to these issues the GAS databases will probably endorse a more detailed, semantic documentation of the results using standardized ontologies even at the clinical level, and as a new element, a detailed, formal multivariate description of the results of the data analysis expressing the remaining uncertainty normatively. Thus various completely standardized, multivariate descriptions with model-based probabilistic characterization can be expected to emerge, which summarize the findings of a study. The discussed set of relevant variables and their interactions (i.e., the MBS set and the MBG submodel) can be proper bases for such a characterization.

From the aspect of researchers, these developments will further blur the border between study design and data analysis, between the "analysis" of the electronically available knowledge and new data sets, and in general between the "knowledge" world and "data" world (Gerstein and Junker 2001). The discussed platform follows this fusion of these worlds as follows.

The model-based information retrieval tool described in Sect. 3 allows the use of the results of Bayesian network-based data analysis in formulating complex queries. Currently, this system is applied over PubMed entries, but we are working on its generalization to work on the top of an arbitrary web-search engine (i.e., to provide model-based information retrieval over the whole semantic web). Similarly, to support the rapid tracking of the results, we generalize the statistical text-mining tool (described in Sect. 3) to process entities on the world-wide web using a general search engine (e.g., providing co-rankings instead of co-occurrence for the pairs of domain entities) not just PubMed entries.

The SNP set selection system described in Sect. 4 is similarly generalized to follow this fusionist trend: Now it can be used to select or weight SNPs for data analysis as well after (!) the measurement. That is, besides designing a study that uses the results of earlier data analyses, the SNP set selection system can be used to select SNPs for subsequent Bayesian data analysis and to generate priors for their relevance.

The discussed Bayesian multilevel methodology using Bayesian network features together with the sequential option allows a broad vision to understand the power of the data and to interpret the results, which makes it especially applicable for the analysis of data with small sample size. It can also incorporate a wide range of priors (e.g., from our SNP selection system). The advantage of the joint usage of different feature levels is multiple: (1) we can better understand the sufficiency of the data, (2) the necessity and possibility of the set-based and dependency-based multivariate analysis, and (3) it can further improve the communication of uncertain conclusions (i.e., to cope with positive results bias). Previously, the BMLA methodology was applied for the analysis of features for the preoperative diagnosis of ovarian cancer (Antal et al., 2006); currently it is used for the analysis of the genomic background of rheumatoid arthritis and for the analysis of chemical properties of autoantigens.

Finally, the functionality of the Bayesian logic module described in Sect. 6 can be illustrated by the overview of the stages of fusion of data and knowledge:

1. Personal fusion of the results in the library
2. Personal fusion of the results and the electronic literature using general information retrieval systems
3. Personal fusion of the multivariate results of data analysis, the results of model-based information retrieval, and the multivariate results of the analysis of the electronic literature (e.g., see Sect. 3)
4. Fusion of the univariate aggregation of the results of data analysis and manually curated knowledge bases (e.g., see Ingenuity Systems 2007)
5. Multivariate fusion of the results of data analysis, bibliometric analysis, and manually curated knowledge bases (e.g., see Sect. 6)

The appearance of the multivariate integration of the results of data analysis with prior knowledge will induce the appearance of a compact, probabilistic, multivariate characterization of the results of data analysis, which can also be submitted to public repositories. The joint usage of many such characterizations together with standardized descriptions of the clinical and environmental levels – besides the more or less already standardized metabolic, proteomic, and genomic levels – is also possible with the same methodology described in Sect. 6.

In summary, the described platform also indicates new functionalities of large-scale, unified GAS knowledge bases handling the uncertainties properly, that is, expressing the network and degree of scientifically corroborated knowledge in a better way.

Table 3 Ratios and posteriors for SNP types

SNP type	Original ratio	Probability of relevance	Ratio of relevance
Intron	0.456522	1.63E−01	6.19E−01
cds-Nonsynon	0.086957	1.00E−01	5.23E−02
UTR-3	0.065217	2.53E−02	9.50E−03
cds-Synon	0.043478	4.43E−04	1.04E−04
UTR-5	0.021739	0.00E+00	0.00E+00
Near gene-3	0.021739	0.00E+00	0.00E+00
<Unknown>	0.304348		

The *Original ratio* column presents the ratio of the types of the measured SNPs. *Probability of relevance* column presents the conditional probability of relevance for a given SNP type (i.e., *P(XOMBS(Asthma)|Type(X) = type)*). The *Ratio of relevance* column presents the ratio of types of relevant SNPs (i.e., *P(Type(X) = type|XOMBS(Asthma))*)

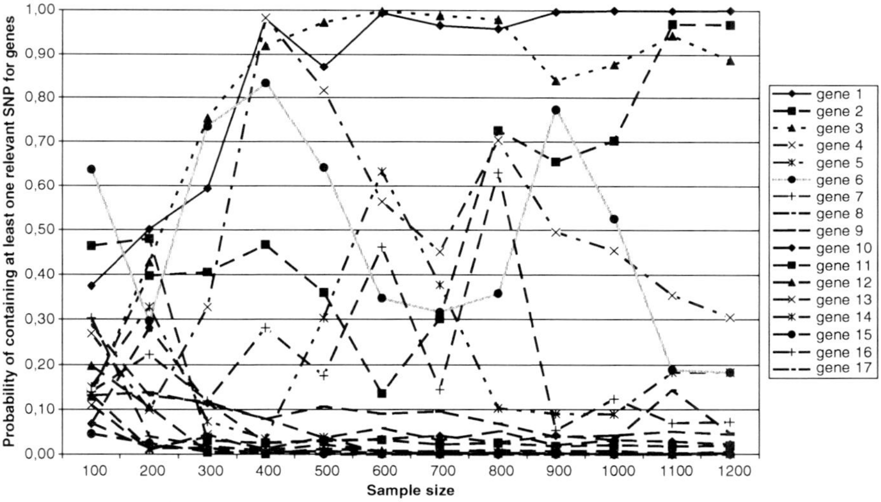

Fig. 6 The sequential posteriors – inferred from growing amount of data – that a given gene contains a SNP relevant for asthma. The probability of relevance is induced by the posterior *p(MBM(SNPi,Asthma|D))*

Acknowledgments We thank Yves Moreau for his insightful suggestion to apply the SNP study design system for prior generation in our Bayesian data analysis. Supported by grants from the OTKA National Scientific Research Fund (PD-76348); NKTH TECH_08-A1/2-2008-0120 (Genagrid), and the János Bolyai Research Scholarship of the Hungarian Academy of Sciences (P. Antal).

References

Aerts S et al (2006) Gene prioritization through genomic data fusion. Nature 24:537–544
Ananiadou S, Mcnaught J (2006) Text mining for biology and biomedicine, Artech House
Antal P, Millinghoffer (2006) A literature mining using Bayesian networks. In Proceedings of third European workshop on probabilistic graphical models, Prague, pp 17–24

Antal P, Fannes G, Moreau Y, Timmerman D, DeMoor B (2004) Using literature and data to learn Bayesian networks as clinical models of ovarian tumors. Artif Intell Med 30:257–281
Antal P, Gezsi A, Hullam G, Millinghoffer A (2006) Learning complex Bayesian network features for classification. In: Proceedings of third European workshop on probabilistic graphical models, Prague, pp 9–16
Balding DJ (2006) A tutorial on statistical methods for population association studies. Nat Rev Genet 7:781–791
Beckman Coulter – SNPStream: http://www.beckmancoulter.com/products/instrument/geneticanalysis/ceq/genomelab_snpstream_dcr.asp
Bonis J et al (2006) OSIRIS: A tool for retrieving literature about sequence variants. Bioinformatics 22(20):2567–2569
Boutilier C, Friedman N, Goldszmidt M, Koller D (1996) Context-Specific Independence in Bayesian Networks, Proc. of the 20th Conf. on Uncertainty in Artificial Intelligence ({UAI}-1996), 115–123
Cooper GF, Herskovits E (1992) A Bayesian method for the induction of probabilistic networks from data. Machine Learning, 9:309–347
Couzin J (2008) MicroRNAs make big impression in disease after disease. Science 319:1782–1784
De La Vega FM et al (2006) A tool for selecting SNPs for association studies based on observed linkage disequilibrium patterns. Pac Symp Biocomput 11:487–498
Denison DGT, Holmes CC, Mallick BK, Smith AFM (2002) Bayesian Methods for Nonlinear Classification and Regression. Wiley & Sons
Estivill X, Armengol L (2007) Copy number variants and common disorders: Filling the gaps and exploring complexity in genome-wide association studies. PLoS Genet 3:1787–1799
Franke A et al (2006) Genomizer: An integrated analysis system for genome wide association data. Hum Mutat 27(6):583–588
Friedman N (2003) Inferring cellular networks using probabilistic graphical models. Science 303(5659):799–805
Friedman N, Koller D (2003) Being Bayesian about network structure. Mach Learn 50(2):95–125
Gamerman D (1997) Markov Chain Monte Carlo. Chapman & Hall, London
Gelman A, Carlin JB, Stern HS, Rubin DB (1995) Bayesian data analysis. Chapman & Hall, London
Gerstein M, Junker J (2001) Blurring the boundaries between scientific "papers" and biological databases. Nature (web debate, on-line 7 May 2001)
Giudici P, Castelo R (2003) Improving Markov Chain Monte Carlo model search for data mining. Machine Learning, 50:127–158
Grover D et al (2007) QuickSNP: An automated web server for selection of tagSNPs. Nucleic Acids Res 35:W115–W120
Gu S et al (2005) HAPLOT: A graphical comparison of haplotype blocks, tagSNP sets and SNP variation for multiple populations. Bioinformatics 21(20):3938–3939
Ingenuity Systems (2007) Ingenuity pathways analysis
Kohavi R, John GH (1997) Wrappers for feature subset selection. Artif Intell 97:273–324
Moffatt MF et al (2007) Genetic variants regulating ORMDL3 expression contribute to the risk of childhood asthma. Nature 448:470–473
Pearl J (1988) Probabilistic reasoning in intelligent systems. Morgan Kaufmann, San Francisco
Peer D, Regev A, Elidan G, Friedman N (2001) Inferring subnetworks from perturbed expression profiles. Bioinformatics, Proc. of ISMB, 17(Suppl. 1):215–224
Petretto E, Liu ET, Aitman TJ (2007) A gene harvest revealing the archeology and complexity of human disease. Nat Genet 39:1299–1301
Pettersson F et al (2004) GOLDsurfer: Three dimensional display of linkage disequilibrium. Bioinformatics 20(17):3241–3243
Russel S, Norvig P (2001) Artificial intelligence. Prentice Hall
Shriner D, Vaughan LK, Padilla MA, Tiwari HK (2007) Problems with genome-wide association studies. Science 316:1840–1842
Szalai C, Ungvári I, Pelyhe L, Tölgyesi G, Falus A (2008) Asthma from a pharmacogenomic point of view. Br J Pharmacol 153:1602–1614

Wang L et al (2005) SNPHunter a bioinformatic software for single nucleotide polymorphism data acquisition and management. BMC Bioinformatics 6:16
Xu H et al (2005) SNPselector: A web tool for selecting SNPs for genetic association studies. Bioinformatics 21(22):4181–4186
Yue P et al (2006) SNPs3D: Candidate gene and SNP selection for association studies. BMC Bioinformatics 7:166
Zhang Y, Liu JS (2007) Bayesian inference of epistatic interactions incase-control studies. Nat Genet 39(9):1167–1173

Index